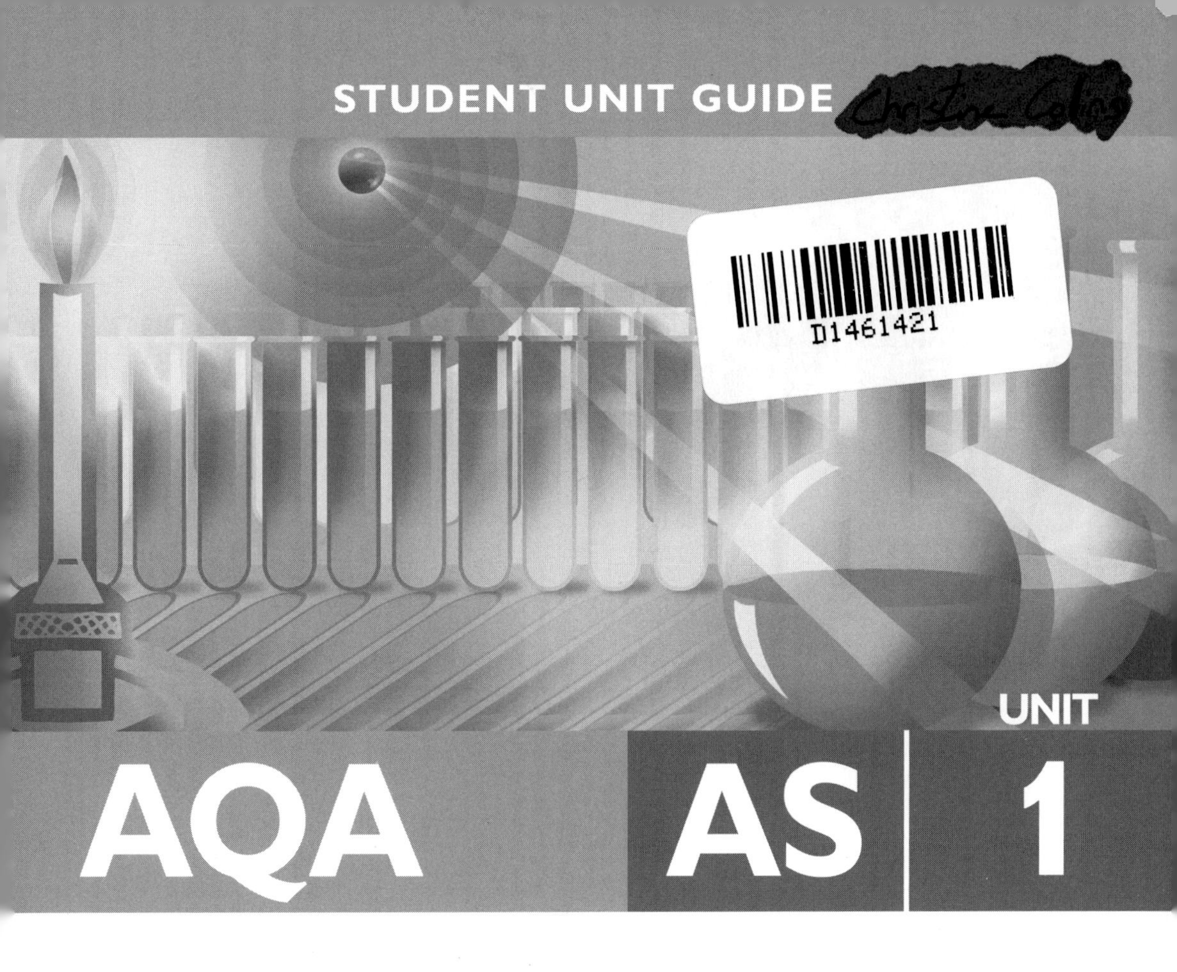


# Chemistry

## Foundation Chemistry

Margaret Cross

I would like to acknowledge the original work of Neil Goldie, who wrote the AQA AS Chemistry Unit 1 Guide for the previous specification. I have drawn extensively from his publication and I am grateful to Alison Goldie for giving me permission to adapt his work for this guide to the new specification.

Philip Allan Updates, an imprint of Hodder Education, an Hachette UK Company, Market Place, Deddington, Oxfordshire OX15 0SE

***Orders***

Bookpoint Ltd, 130 Milton Park, Abingdon, Oxfordshire OX14 4SB
tel: 01235 827720
fax: 01235 400454
e-mail: uk.orders@bookpoint.co.uk
Lines are open 9.00 a.m.–5.00 p.m., Monday to Saturday, with a 24-hour message answering service. You can also order through the Philip Allan Updates website: www.philipallan.co.uk



ISBN 978-0-340-97110-9

First printed 2008
Impression number 5 4 3 2
Year 2013 2012 2011 2010 2009



Typeset by Fakenham Photosetting, Fakenham, Norfolk
Printed by MPG Books, Bodmin

# Contents

# Introduction

## About this guide

This guide is for students following the AQA Chemistry AS specification. It deals with **Unit 1: Foundation Chemistry**. This unit covers $33\frac{1}{3}\%$ of the total AS marks, or $16\frac{2}{3}\%$ of the total A-level marks. The examination lasts for 1 hour 15 minutes and consists of four or five short-answer structured questions and one longer structured question that requires the answer to be written in continuous prose. All questions are compulsory. There are 70 marks available on this paper.

Unit 1 is probably the most important, because success in all future modules depends on an understanding of the basic principles and concepts covered here.

### The key to success

It is essential that you can recall the basic facts and definitions, but a deeper understanding of the subject is necessary if you want to achieve the top grade. The key to success in chemistry is to understand the fundamental concepts and be able to apply them to new and unfamiliar situations. Good examination technique is also an important factor, enabling you to work more effectively in the exam and access the marks needed for a top grade.

This guide allows you to look again at the content of the specification, to test yourself at the end of each section and to assess your own work. Reading through the examiner's comments will help you to improve your exam technique. Once you have worked through this guide, you will be aware of your weak areas and it is these areas that you need to address. Make a list of these weak areas and discuss any problems with other students in your class and with your teacher.

### Using this guide

This guide has three sections:

- **Introduction** — this provides guidance on study and revision, together with advice on approaches and techniques to ensure you answer the examination questions in the best way that you can.
- **Content Guidance** — this section is not intended to be a textbook. It offers guidelines on the main features of the content of Unit 1, together with particular advice on making study more productive.
- **Questions and Answers** — this shows you the sort of questions you can expect in the unit test. Grade-A answers are provided; these are followed by examiner's comments. Careful consideration of these will improve your answers and, much more importantly, will improve your understanding of the chemistry involved.

When you have revised a topic using the Content Guidance section, if there is something that you do not understand, you should refer to your class notes and textbooks.

Write down specific questions and discuss them with your teacher. For instance, 'Please could you explain bonding again' will not lead to good use of your time if you understand most of the features of ionic, covalent and metallic bonds. 'Please could you explain what is meant by the term coordinate bond and give me some examples' is more specific and shows that you have worked hard to identify weak areas.

Once you have revised a particular topic thoroughly, you should attempt the relevant question in the Questions and Answers section, *without* looking at the grade-A answer.

Compare your answer with the grade-A answer and estimate your own performance. A rough guide to use is 80% = grade A, 70% = grade B, 60% = grade C. However, these grade boundaries vary, depending on the individual paper and the performance of the candidates.

Read through the examiner's comments to find out if you have made any of the common mistakes and to see how you could improve your technique. The comments also give some alternative answers.

Make a note of *specific questions* that caused you problems and discuss them with other students and your teacher.

# Revision schedule

- Plan your revision schedule carefully.
- Revise regularly.
- Leave yourself enough time to cover all the material. You need to go through each topic once as a basic minimum and then go through the weak areas again.
- Do not try to achieve too much in each revision session. Revise one topic per session, e.g. ionisation energies. Here is one way to structure your session:
  - revise from the Content Guidance section (and your own notes)
  - make a brief written summary (no more than an A4 sheet of paper)
  - attempt the question
  - mark your answer
  - read the examiner's comments
- If you have scored a grade A, then tick the topic in the revision checklist table below. You must have a break before you start the revision of the next topic.
- If there are weak areas and questions that you clearly do not understand, then write down specific questions ready for discussion with your teacher.
- Finally, make sure that you attempt some recent past paper questions from the exam board and study the mark schemes carefully. Although identical questions are unlikely, similar questions may be set.

## Revision checklist

Once you have completed your revision and feel that you understand the topic fully, tick the relevant box. If you have identified a weak area, place a cross in the relevant box.

### Atomic structure

| Topic | Details | ✓ or ✗ |
|---|---|---|
| Structure of atoms | Protons, neutrons, electrons<br>Mass number and atomic number<br>Isotopes | |
| Mass spectrometry | Basic principles of operation<br>Interpretation of spectra; calculation of $A_r$<br>Relative molecular mass | |
| Electron arrangement | *s*, *p*, *d* levels for atoms and ions up to atomic number 36 | |
| Ionisation energies | Definition<br>Variation across period 3<br>Variation down group 2<br>Successive ionisation energies | |

### Amount of substance

| Topic | Details | ✓ or ✗ |
|---|---|---|
| Relative $A_r$ and $M_r$ | Definitions | |
| The mole and the Avogadro constant | Concept as applied to various particles<br>Formulae and equations<br>Calculations using $n = m/M_r$ | |
| Ideal gas equation | Recall the equation<br>Use it to calculate $M_r$ | |
| Empirical and molecular formulae | Definitions<br>Calculation of empirical formula<br>Deduction of molecular formula | |
| Equations and associated calculations | Writing equations for reactions<br>Reacting masses<br>Reacting volumes<br>Titration calculations<br>% yield; % atom economy | |

## Bonding

| Topic | Details | ✓ or ✗ |
|---|---|---|
| Types of bonding | Ionic<br>Covalent<br>Dative covalent (coordinate)<br>Metallic<br>Bond polarity | |
| Intermolecular forces | Induced dipole–dipole (van der Waals) forces<br>Permanent dipole–dipole forces<br>Hydrogen bonding | |
| States of matter | Energy changes associated with changes of state | |
| Types of crystal | Ionic<br>Metallic<br>Molecular<br>Macromolecular | |
| Shapes of molecules | Basic principles and examples | |

## Periodicity

| Topic | Details | ✓ or ✗ |
|---|---|---|
| Classifying elements | *s*, *p*, *d* block elements | |
| Physical properties of period 3 elements | Trends and reasons for trends in:<br>• Atomic radius<br>• First ionisation energies<br>• Melting and boiling points | |

## Introduction to organic chemistry

| Topic | Details | ✓ or ✗ |
|---|---|---|
| Nomenclature | Empirical formula<br>Molecular and structural formulae<br>Homologous series and functional groups<br>Naming of simple organic compounds | |
| Isomerism | Structural isomerism<br>Chain and position isomerism | |

**Alkanes**

| Topic | Details | ✓ or ✗ |
|---|---|---|
| Fractional distillation of crude oil | Petroleum and alkanes<br>Separation of the fractions | |
| Cracking of alkanes | Definition<br>Economic reasons for cracking alkanes<br>Thermal cracking<br>Catalytic cracking | |
| Combustion of hydrocarbons | Complete and incomplete combustion<br>Formation of pollutants and their removal by catalytic converters<br>Formation of sulfur dioxide<br>Combustion of fossil fuels and release of carbon dioxide into the atmosphere<br>Greenhouse gases and their contribution to global warming | |

# Unit Test 1

If you have revised thoroughly, completed all the questions in this guide and discussed problems with other students and your teacher, you should enjoy the exam. If you have completed AQA past papers, then the style of the paper will be familiar and you will recognise some questions in the exam because they will be similar to previous questions.

Do not begin to write as soon as you open the paper — quickly scan the questions first.

It is *not* essential that you answer the questions in order. If the first question is difficult, then leave it to the end. It *is* essential that you answer *all* the questions.

You will have enough time to answer all the questions, provided you keep your answers concise and do not include irrelevant information. It is easy to waste time writing out a section of your notes that is irrelevant to the question asked. Do not repeat the question when starting your answer. The key to exam success is achieving the maximum number of marks in the minimum number of words.

The mark allocation at the end of each question should be used to estimate the amount of detail needed in your answer. If there is 1 mark available, the examiner is looking for a key word or phrase and certainly no more than one sentence. If there are 4 marks available, you should include four key points, which usually means writing four short sentences.

No marks are available for producing neat answers, but it certainly helps the examiners when they are marking your work. Untidy diagrams may become inaccurate and this definitely loses marks.

## Terms used in examination questions

### Give/state/name

These appear in many of the early structured questions. You need only write one or two words. There is usually 1 mark available for questions of this type.

### Define

It is essential that you learn all your definitions. You need to state definitions and laws concisely. Definitions are usually worth 2 or 3 marks and the mark allocation indicates how many key points must be included.

### Complete

You have to finish off a diagram, a graph or a table.

### Draw the structure

When drawing the structures of organic compounds it is essential that each carbon atom is surrounded by four bonds. You must include all the hydrogen atoms in the structure.

### Draw/show by means of a diagram

Most drawings require only a simple sketch. It must be clear and accurate, but the examiner does not expect it to be of the same standard as that seen in your textbook. Poor presentation can make diagrams inaccurate and this is when marks are lost.

### Calculate or determine

Use information to calculate a final answer that must be shown to the correct number of significant figures. Always show your working and include appropriate units after the final answer.

### Write a balanced equation

In chemical equations, state symbols are not usually expected unless the examiner specifically asks for them. In physical processes and thermochemistry equations, state symbols are usually expected. For example, the first ionisation energy of sodium is:

$$Na(g) \rightarrow Na^+(g) + e^-$$

In organic reactions it is important to show the structure of the reactants and the products rather than using molecular formulae.

### Describe/explain

'Describe' means only give a description, whereas 'explain' requires a reason or interpretation. Both these terms mean more depth is required in your answer. You can judge how much detail is required by the mark allocation. 10 marks means that ten key points must be covered and this will probably require you to write a minimum of ten sentences.

### Use

Questions using this term often include some data. Make sure that you do use the data and include them in your answer. You may have to use a basic principle and apply it to a new and unfamiliar situation.

### Suggest

This means that you will probably not have come across the material in the question before. You will have been taught the basic principle and now you have to apply this knowledge to an unfamiliar situation.

# Content Guidance

This section covers the content of **Unit 1: Foundation Chemistry**. It is the most important unit at AS and A-level because it covers many basic concepts that must be fully understood to guarantee success in all units.

In this Content Guidance section, the specification has been converted into user-friendly language and is in a format that is easy to remember. All the key facts, definitions and basic principles are covered. In order to achieve a top grade, it is essential that you fully understand the basic concepts and that you can apply them to unfamiliar situations. The content of this unit falls into six sections, which are summarised below.

**Atomic structure:** Much of this has been covered at GCSE. The new topics are ionisation energies and the mass spectrometer, which are regularly tested. These new topics appear quite demanding at the beginning of the course. However, questions on these topics are usually well answered. Electron arrangements are now in terms of sub-levels.

**Amount of substance:** This topic is usually covered in the early stages of the AS course. Many students find the concepts difficult to grasp and the topic difficult to revise. The only way to revise calculations effectively is to practise examples from past papers and textbooks — repeating the process makes you more confident. Sometimes the calculation will still be too difficult, but questions on this topic also require recall of some basic definitions, giving you access to some of the marks.

**Bonding:** This builds on many of the basic concepts covered at GCSE. However, dative covalency, polarisation and different types of intermolecular force are all new concepts. This section can be quite demanding because of all the new ideas, so it is essential to simplify the material. 'Shapes of molecules' is a new topic that most candidates find easy, if taught in a systematic way.

**Periodicity:** This is a good topic to test your understanding of atomic structure and bonding. If you understand the basic concepts of structure and bonding, it is easy to explain the observed trends in the physical properties of the elements of period 3.

**Introduction to organic chemistry:** Questions on this topic might ask you to recall definitions of empirical, molecular and structural formulae, homologous series and functional groups, or to name some simple organic compounds. You need to understand the concepts of structural isomerism (chain and position).

**Alkanes:** You may be asked to describe the separation of petroleum into fractions. You need to understand the principles of fractional distillation and be able to recall the names and uses of the fractions. The cracking of larger fractions into more useful smaller fractions and alkenes and the equations for these reactions are covered in this unit. The different pollutants formed in the internal combustion engine, how they are formed, their effects on the environment and how they are removed by a catalytic converter are also covered.

# Atomic structure

## Fundamental particles: protons, neutrons and electrons

This first topic builds on the basic ideas covered at GCSE. Questions on this topic are relatively straightforward and are usually worth only a few marks in the early structured questions.

The properties of the **fundamental particles** are summarised in the table below.

| Particle | Relative charge | Relative mass |
|---|---|---|
| Proton | +1 | 1 |
| Neutron | 0 | 1 |
| Electron | −1 | 1/1840 |

The relative mass of an electron is so small that, in answer to a question about it, *negligible* or *zero* are acceptable.

In one model of atomic structure, an atom consists of **electrons** (e) surrounding a nucleus that contains **protons** (p) and **neutrons** (n).

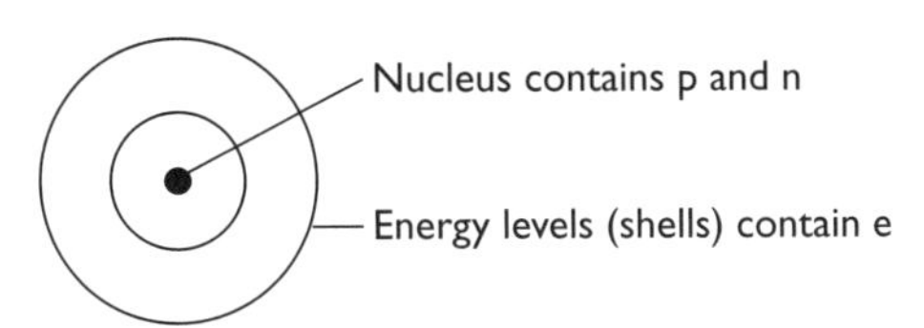


### Mass number and atomic number

The symbols of the elements give information on the **atomic number** and the **mass number**, which in turn reveal the number of fundamental particles in each atom of the element.

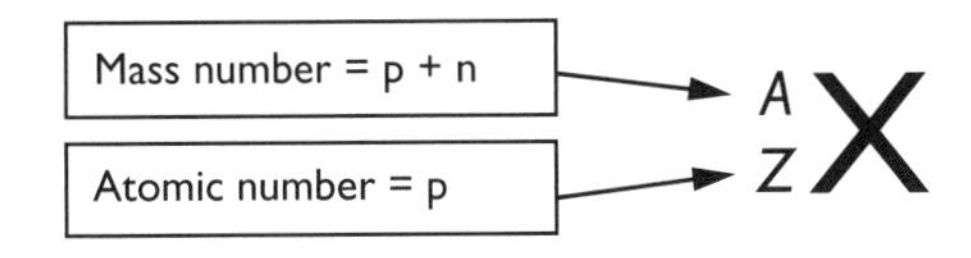


The number of electrons in an atom equals the number of protons.

The number of protons and neutrons never changes in ion formation, but the number of electrons does.

This negative ion has two additional electrons

$^{A}_{Z}X^{3+}$

This positive ion has lost three electrons

**Example**

State the number of fundamental particles in the following species: $^{31}_{15}P$; $^{24}_{12}Mg^{2+}$; $^{35}_{17}Cl^{-}$.

**Answer**

| Species | Protons | Electrons | Neutrons |
|---|---|---|---|
| $^{31}_{15}P$ | 15 | 15 | 16 |
| $^{24}_{12}Mg^{2+}$ | 12 | 10 | 12 |
| $^{35}_{17}Cl^{-}$ | 17 | 18 | 18 |

## Isotopes

- Isotopes are the same element with the same atomic number but a different mass number.
- Isotopes have the same number of protons and the same number of electrons, but a different number of neutrons.

The isotopes of magnesium can be used to illustrate all the basic principles. The three isotopes of magnesium are shown in the table below, with the number of protons, electrons and neutrons listed for each isotope.

| $^{24}_{12}Mg$ | $^{25}_{12}Mg$ | $^{26}_{12}Mg$ |
|---|---|---|
| 12 protons | 12 protons | 12 protons |
| 12 electrons | 12 electrons | 12 electrons |
| 12 neutrons | **13** neutrons | **14** neutrons |

Other elements could be used to illustrate these principles.

## The mass spectrometer

The principles of a simple mass spectrometer are limited to ionisation, acceleration, deflection and detection.

The apparatus illustrated can be used to separate the three isotopes of magnesium. If a diagram is required in the unit test, then a simple sketch is sufficient. If a diagram is provided, you should be able to label it, particularly the paths of the different ions. In the same strength of magnetic field, the ion with the smallest mass-to-charge ratio is deflected the most.

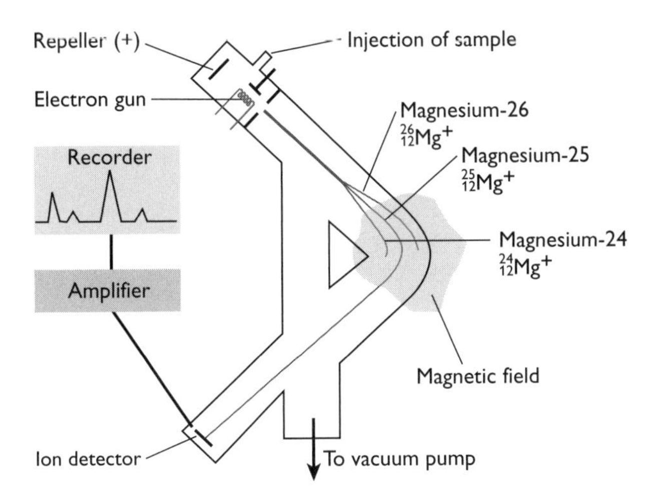


An explanation of how the mass spectrometer is used to separate isotopes is a common question. The most common mistake is the inclusion of too much detail. There are four key stages in the separation of the isotopes and you need to mention four key points for each stage. The stages are ionisation, acceleration, deflection and detection.

**Ionisation**

- The sample is vaporised.
- It is bombarded with electrons using an electron gun.
- This knocks out an electron from each atom (or molecule) of the sample.
- Positive ions (cations) are produced.

**Acceleration**

- The cations are accelerated.
- Acceleration is achieved by passing the cations through an electric field.
- The ions move towards a negative plate.
- The ions are focused into a beam by slits.

**Deflection**

- The ions are deflected.
- Deflection is achieved by passing the ions through a magnetic field.
- Deflection depends on the mass-to-charge ratio ($m/z$).
- The smaller the $m/z$ ratio, the greater is the deflection.

**Detection**

- The ions hit a detector.
- Current is produced.
- Information is fed to a computer or displayed on a chart.
- Varying either the electric field or magnetic field collects ions of different masses.

## Interpretation of simple mass spectra of elements

The mass spectrum of magnesium is shown below.

- It has three peaks, confirming that three isotopes are present.
- The peak at $m/z$ value 24 has the highest relative intensity, so this means it is the most abundant isotope.

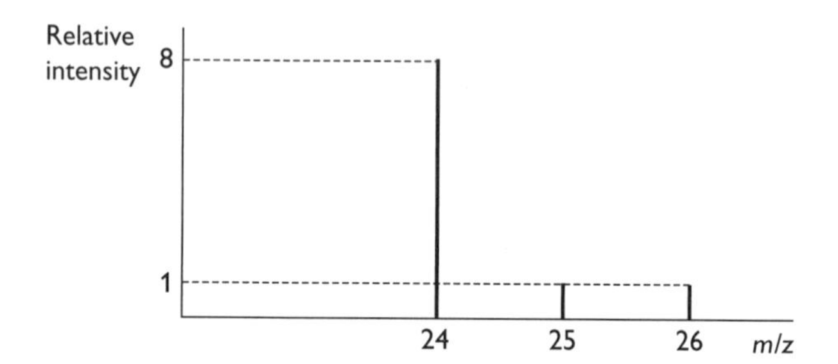


## Calculation of relative atomic mass from isotopic abundance

This is limited to mononuclear ions.

### Example 1: the relative atomic mass of magnesium

Using the mass spectrum of magnesium, determine the average relative atomic mass ($A_r$).

**Answer**

$$A_r = \frac{(8 \times 24) + (1 \times 25) + (1 \times 26)}{10} = 24.3$$

***Tip*** Data for use in these calculations can be presented in different ways. Careless mistakes are often made by using the incorrect denominator in the calculation — this should be the sum of the (relative) intensities or 100 if percentage abundances are given.

### Example 2: the relative atomic mass of boron

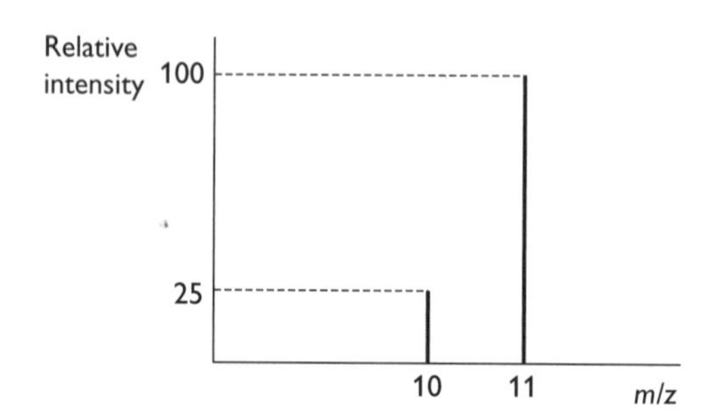

Determine the $A_r$ of boron.

**Answer**

$$A_r = \frac{(25 \times 10) + (100 \times 11)}{125} = 10.8$$

***Tip*** A common mistake is to use an incorrect denominator, thus:

$$A_r = \frac{(25 \times 10) + (100 \times 11)}{100} = 13.5$$

**Example 3: identifying an unknown element**

Use the data in the table to identify the element.

| *m/z* | 80 | 82 | 83 | 84 | 86 |
|---|---|---|---|---|---|
| **Relative intensity** | 1 | 5 | 5 | 25 | 8 |

**Answer**

$$A_r = \frac{(80 \times 1) + (82 \times 5) + (83 \times 5) + (84 \times 25) + (86 \times 8)}{\text{total intensity of 44}} = 83.9$$

The answer is krypton.

***Tip*** A common mistake is to give the answer as polonium. Remember, when identifying an element from mass spectra data, you must look at mass numbers, *not* atomic numbers.

## Use of mass spectrometry to determine relative molecular mass

The spectra of compounds can be complicated when they break up and produce smaller ions. The $M_r$ of the compound can be determined from spectra by looking at the peak with the largest *m/z* value. (This is covered in more detail in Unit 2).

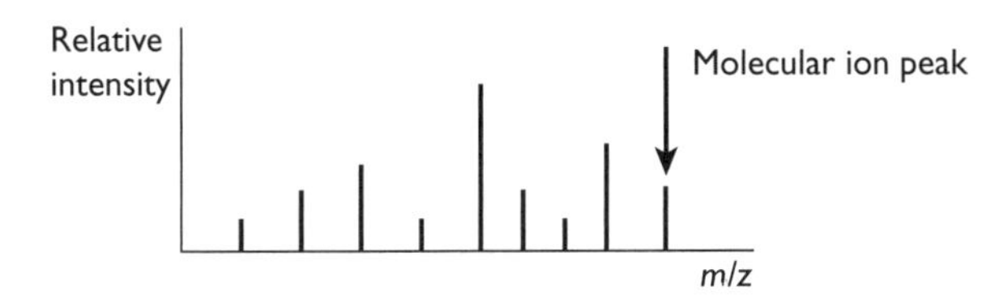


The final peak is the molecular ion peak. It is equivalent to the relative molecular mass.

# Electron arrangements

Few questions are set on electron arrangements. However, it is essential that the basic principles are understood because they are needed to explain many other topics.

Consider the energy diagram below for the first 36 electrons.

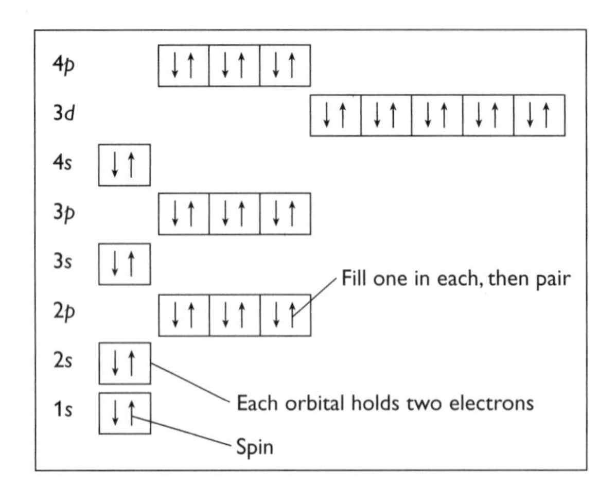


Refer to this diagram when reading through the following key points:

- Electrons are arranged in **energy levels** (shells) from level 1 (lowest energy) to level 4 (highest energy).
- Each energy level contains **sub-levels** designated *s*, *p*, *d* (and *f*).
- Level 1 contains one sub-level (*s*), level 2 contains two sub-levels (*s* and *p*), level 3 contains three sub-levels (*s*, *p* and *d*), level 4 contains four sub-levels (*s*, *p*, *d* and *f*).
- Each sub-level contains a different number of **orbitals** and each orbital can hold a maximum of two electrons.
- An ***s* sub-level** contains one orbital and so can hold two electrons; a ***p* sub-level** contains three orbitals and so can hold six electrons; a ***d* sub-level** contains five orbitals and so can hold ten electrons.
- When filling up the energy diagram above, remember that electrons enter the lowest energy levels first.
- When filling a 2*p* sub-level, three electrons are first placed in separate orbitals [↑] [↑] [↑] so that they are **unpaired**. Then the next three electrons are added to complete the sub-level [↑↓] [↑↓] [↑↓]. This arrangement is also followed when filling the 3*d* sub-level.
- The main complication is the electron arrangements of the transition metal atoms and their ions. Looking at the energy diagram above it is clear that the 4*s* level is filled before the 3*d*. However, when forming ions, 4*s* electrons are lost before 3*d* electrons.

- Two transition elements have unusual electron configurations. These are chromium and copper:

  Cr $1s^2\,2s^2\,2p^6\,3s^2\,3p^6\,3d^5\,4s^1$

  Cu $1s^2\,2s^2\,2p^6\,3s^2\,3p^6\,3d^{10}\,4s^1$

The examples in the table below show the electron configurations of some common atoms and ions. The number of electrons in each sub-level is always written as a superscript, for example $1s^2$, not 1s2.

| | |
|---|---|
| $_{15}P$ | $1s^2\,2s^2\,2p^6\,3s^2\,3p^3$ |
| $_{19}K$ | $1s^2\,2s^2\,2p^6\,3s^2\,3p^6\,4s^1$ |
| $_{30}Zn$ | $1s^2\,2s^2\,2p^6\,3s^2\,3p^6\,4s^2\,3d^{10}$ (usually written $1s^2\,2s^2\,2p^6\,3s^2\,3p^6\,3d^{10}\,4s^2$) |
| $_{26}Fe$ | $1s^2\,2s^2\,2p^6\,3s^2\,3p^6\,3d^6\,4s^2$ |
| $_{26}Fe^{2+}$ | $1s^2\,2s^2\,2p^6\,3s^2\,3p^6\,3d^6$ |
| $_{26}Fe^{3+}$ | $1s^2\,2s^2\,2p^6\,3s^2\,3p^6\,3d^5$ |

## Ionisation energies

The definition of the first ionisation energy — $\mathbf{X(g) \rightarrow X^+(g) + e^-}$ — covers three key points. It is the energy required to convert **1 mole** of **gaseous atoms** into 1 mole of **gaseous cations**, each atom losing an electron.

### Variation across the periodic table

The graphs presented in many textbooks look complicated. Keep your graph simple and remember the pattern '2–3–3, then big drop'. This 2–3–3 pattern is the same for both period 2 and period 3.

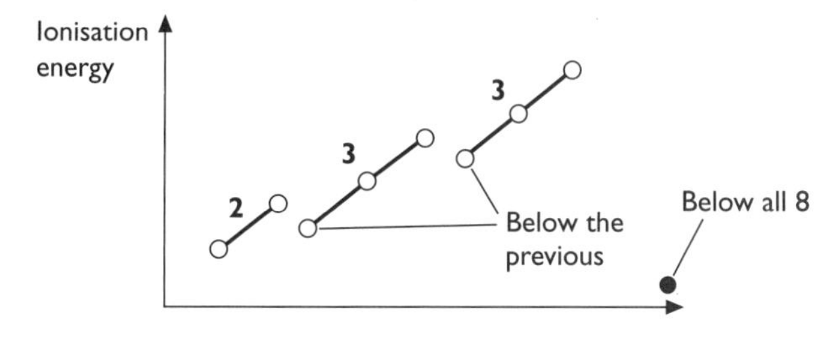


Join the points up to complete the graph. This shows the variation across period 2 or period 3.

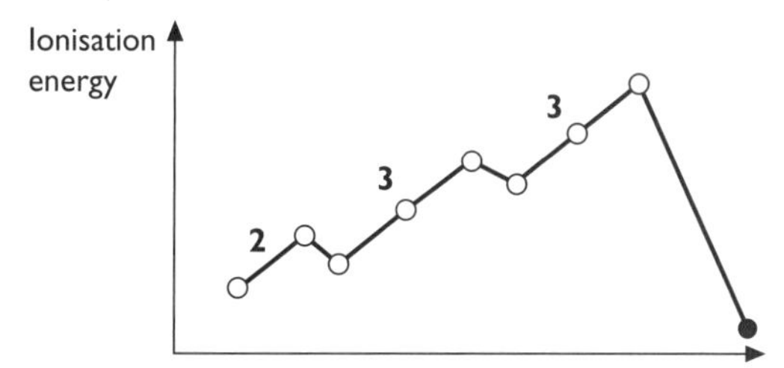

You could be asked to explain the following features of the graph:

- the general increase across the period
- the drops between groups 2 and 3 and between groups 5 and 6
- the big drop at the end of a period

These are common exam questions.

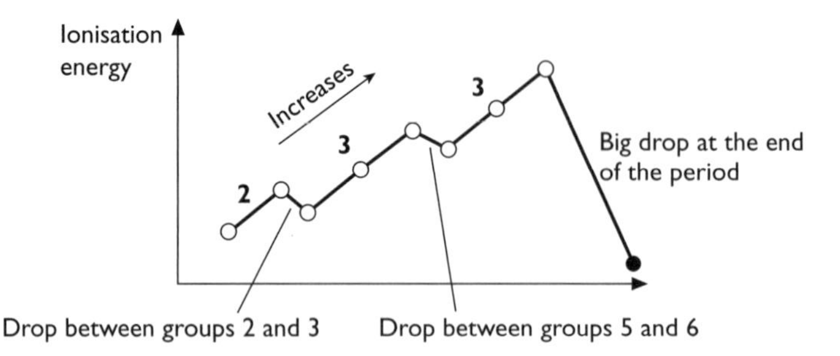


***General increase across the period***

Key points are:

- increase in proton number (or nuclear charge)
- shielding the same (or electrons enter the same energy level)
- increased attraction for electrons

***Drop between groups 2 and 3***

Drawing the electron arrangements makes it easier to score the marks:

Mg: $1s^2\,2s^2\,2p^6\,3s^2$

Al: $1s^2\,2s^2\,2p^6\,3s^2\,3p^1$

- The electron is removed from the **3*p* level**.
- 3*p* is higher in energy than 3*s*, so the electron is easier to remove.

The same argument can be used for beryllium and boron, but the electron is removed from the 2*p* level, which is higher in energy than the 2*s*.

***Drop between groups 5 and 6***

Draw the electron arrangements of the two atoms and emphasise the arrangement of the electrons in the 3*p* level:

P: $1s^2\,2s^2\,2p^6\,3s^2\,3p^3$ [↑] [↑] [↑]

S: $1s^2\,2s^2\,2p^6\,3s^2\,3p^4$ [↑↓] [↑] [↑]

- There are **paired electrons** in one of the 3*p* orbitals in sulfur.
- This causes **repulsion**, so the electron is easier to remove.

The same argument can be used for nitrogen and oxygen, but the electrons are paired in one of the 2*p* orbitals.

***The big drop at the end of a period***

This is the same explanation as for the decrease in ionisation energies down any group:

- more energy levels (or increase in atomic radii)
- more shielding
- less attraction for outer electrons

### Successive ionisation energies

Second ionisation energy:

$$X^{+}(g) \rightarrow X^{2+}(g) + e^{-}$$

Third ionisation energy:

$$X^{2+}(g) \rightarrow X^{3+}(g) + e^{-}$$

Successive ionisation energies allow us to deduce how many electrons are in the outside energy level of an element.

- Successive ionisation energies increase — the same number of protons and fewer electrons mean more attraction for the remaining electrons.
- Outer electrons are easier to remove because they are shielded by the inner electrons.
- The big jump in ionisation energy indicates the electron is removed from a new inner energy level, which is closer to the nucleus and less shielded by other electrons.

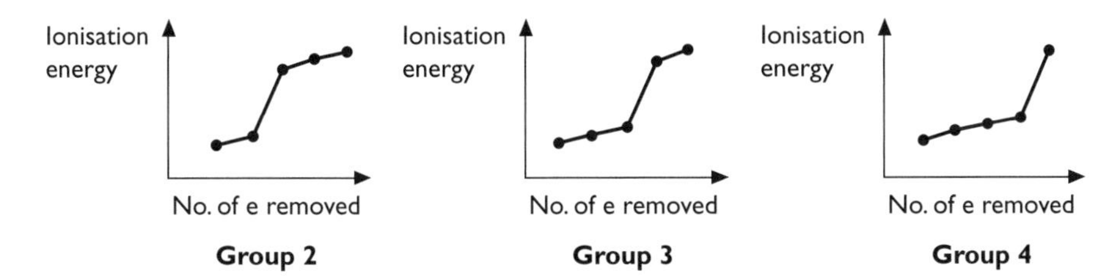


# Amount of substance

The chemist's unit of the **amount of substance** is the number of **moles** (**mol**) of the substance. You need to understand how the mole is defined and to be able to calculate the number of moles of a substance from data provided. This also involves understanding and learning some other definitions and relationships.

Calculations involving the use of moles form the basis of many of the calculations that you will meet in later units. Therefore, it is essential that for this unit you practise as many different types of calculation as possible from books, past papers and your class notes. You should set out your calculations clearly so that your working can be followed easily.

The questions involving mole calculations are very structured at AS to guide you through to the final answer. If you make a mistake at any stage in the calculation and obtain the wrong answer, you can still score the marks for the remaining stages, provided your working is shown clearly and no further mistakes are made.

# Relative masses, the mole and the Avogadro constant

## Definitions

- **Relative atomic mass**, $A_r$, is the mass of one atom of an element relative to $\frac{1}{12}$ the mass of one atom of carbon-12, which has a mass of 12.00 atomic mass units.

$$A_r = \frac{\text{mass of one atom of an element}}{\frac{1}{12}\text{ mass of one atom of }^{12}\text{C}}$$

- The **relative molecular mass**, $M_r$, of a substance is the mass of one molecule of the substance relative to $\frac{1}{12}$ the mass of one atom of carbon-12, which has a mass of 12.00 atomic mass units.

$$M_r = \frac{\text{mass of one molecule of the substance}}{\frac{1}{12}\text{ mass of one atom of }^{12}\text{C}}$$

  As only covalently bonded substances exist as molecules, it is more accurate to use the term **relative formula mass** for other substances. This is also represented by $M_r$ and is the sum of all the relative atomic masses of the atoms in the formula of the substance.

*Note*: Relative atomic mass, relative molecular mass and relative formula mass have no units.

- The **mole** is defined as the amount of substance that contains the same number of elementary particles (e.g. atoms, molecules, ions, electrons) as there are atoms in 12 g of carbon-12. It is given the unit **mol**.

*Note*: When mentioning moles, you must always state the type of particles to which you are referring – for example, 1 mole of sodium ions or 1 mole of hydrogen molecules.

- The **Avogadro constant** is the number of particles in 1 mole of any substance. This constant number of particles is $\mathbf{6.023 \times 10^{23}\ mol^{-1}}$. The Avogadro constant is given the symbol **L**.
- The **molar mass** of a substance is the mass, in grams, of 1 mole of the substance. For example, the molar mass of sodium hydroxide, NaOH, is $40.0\ g\ mol^{-1}$.
- The number of moles is related to the mass and the $M_r$ of a substance by the following equation:

$$\textbf{number of moles } (n) = \frac{\textbf{mass (in grams)}}{\textbf{molar mass}}$$

***Tip*** Learn this final equation. In order to do calculations involving moles of substances, you need to be able both to apply it and to rearrange it.

**Example 1**

Calculate the number of moles of oxygen molecules, $O_2$, in 72.0 g of oxygen gas.

**Answer**

$$\text{number of moles (of } O_2 \text{ molecules)} = \frac{\text{mass}}{\text{molar mass}} = \frac{m}{M_r}$$

$$= \frac{72.0}{2 \times 16.0} = 2.25\,\text{mol}$$

**Example 2**

Calculate the mass of 0.151 mol of sodium chloride, NaCl.

**Answer**

mass(g) = number of moles × molar mass

molar mass of NaCl = 23.0 + 35.5 = 58.5 g $mol^{-1}$

mass = 0.151 × 58.5 = 8.8335 = 8.83 g (to 3 sig. fig.)

***Tip*** Answers should be given to the number of significant figures appropriate to those given in the data, which is usually three significant figures. Be careful not to confuse significant figures with decimal places.

## Concentration of solutions

The concentration of a solution is the amount of substance per unit volume of *solution* (*not* per unit volume of solvent). The units of concentration are **mol $dm^{-3}$**. For example, a solution made by dissolving 40.0 g of solid sodium hydroxide ($M_r$ = 40.0) in water and then making the solution up to a total volume of 1 $dm^3$ has a concentration of 1.00 mol $dm^{-3}$.

A solution of exactly known concentration is known as a **standard solution** and is made up in a volumetric flask. It is not necessary to always make up 1 $dm^3$ of solution and volumetric flasks of various volumes are available.

It is important to learn the following relationship and to be able to apply it, rearranged if necessary, when doing calculations involving solutions:

**number of moles** (of solute) = **concentration (mol $dm^{-3}$)** × **volume ($dm^3$)**

As volumes used in a reaction are most often measured in $cm^3$, this expression can be written as:

$$\textbf{number of moles} \text{ (of solute)} = \frac{\textbf{concentration (mol dm}^{-3}\textbf{)} \times \textbf{volume (cm}^{3}\textbf{)}}{\textbf{1000}}$$

or

$$n = \frac{\textbf{concentration} \times V}{\textbf{1000}}$$

**Example 1**

What mass of aminosulfonic acid, $H_2NSO_3H$ ($M_r$ = 97.1), is required to make 250 cm$^3$ of solution of concentration 0.100 mol dm$^{-3}$?

**Answer**

$$\text{number of moles} = \text{concentration} \times \frac{V}{1000}$$

$$= 0.100 \times \frac{250}{1000} = 0.0250\,\text{mol}$$

$$\text{mass} = \text{number of moles} \times M_r$$
$$= 0.0250 \times 97.1 = 2.4275\,\text{g} = 2.43\,\text{g (3 sig. fig.)}$$

**Example 2**

What is the concentration, in mol dm$^{-3}$, of a solution containing 4.00 g of sodium hydroxide, NaOH ($M_r$ = 40.0), in 200 cm$^3$ of solution?

**Answer**

$$\text{number of moles} = \frac{\text{mass}}{M_r}$$

$$= \frac{4.00}{40.0} = 0.100\,\text{mol}$$

$$\text{concentration} = \text{number of moles} \times \frac{1000}{V}$$

$$= 0.100 \times \frac{1000}{200} = 0.500\,\text{mol dm}^{-3}$$

# The ideal gas equation

You need to be able to recall the ideal gas equation and use it to calculate relative molecular masses or to predict, from a given equation, the amount of gas produced. The most common mistake is made when converting the units given in the question. If you cannot attempt the calculation, then quote the equation $pV = nRT$ because it is usually worth 1 mark.

| Symbol | Name | Units | Notes |
|---|---|---|---|
| $p$ | Pressure | Pa or kPa | Atmospheric pressure is 101 kPa, which is $1.01 \times 10^5$ Pa |
| $V$ | Volume | m$^3$ | 1 m$^3$ = $1 \times 10^3$ dm$^3$ = $1 \times 10^6$ cm$^3$<br>1 cm$^3$ = $1 \times 10^{-3}$ dm$^3$ = $1 \times 10^{-6}$ m$^3$ |
| $n$ | Moles | mol | |
| $R$ | Gas constant | J K$^{-1}$ mol$^{-1}$ | This value remains constant at 8.31 J K$^{-1}$ mol$^{-1}$ |
| $T$ | Temperature | K | 0°C = 273 K (if temperature is in °C, then add 273) |

## Calculating the relative molecular mass ($M_r$) of a gas

$$pV = nRT$$

where $n = \dfrac{\text{mass (g)}}{\text{relative molecular mass}} = \dfrac{m}{M_r}$

$$pV = \frac{mRT}{M_r}$$

Therefore:

$$\boldsymbol{M_r = \frac{mRT}{pV}}$$

### Example 1

Calculate the $M_r$ of krypton given that 1.81 g of krypton has a volume of 500 $cm^3$ at a temperature of 0°C and a pressure of $9.80 \times 10^4$ Pa.

**Answer**

Using the above equation for $M_r$, this calculation can be shown in one step:

$$M_r = \frac{1.81 \times 8.31 \times 273}{9.80 \times 10^4 \times 500 \times 10^{-6}} = 83.8$$

### Example 2

What volume of oxygen at 25°C and 101 kPa is evolved from the decomposition of 1.70 g of hydrogen peroxide?

$$2H_2O_2(l) \rightarrow 2H_2O(l) + O_2(g)$$

**Answer**

$$\text{moles of } H_2O_2 = \frac{m}{M_r}$$

$$= \frac{1.70}{34.0} = 0.0500 \text{ mol}$$

Using the 2:1 ratio in the equation, moles of oxygen = 0.0250 mol

Rearranging $pV = nRT$ gives $V = \dfrac{nRT}{p}$

$$= \frac{0.0250 \times 8.31 \times 298}{101 \times 10^3}$$

$$= 6.13 \times 10^{-4}\ m^3 = 0.613\ dm^3$$

# Empirical and molecular formulae

## Definitions

- The **empirical formula** is the *simplest whole number ratio* of the *atoms* of each element in a compound. The empirical formula of hydrogen peroxide is HO.
- The **molecular formula** is the *actual* number of atoms of each element present in the compound. The molecular formula is a simple multiple of the empirical formula. The molecular formula of hydrogen peroxide is $H_2O_2$.

### Example 1

A chloride of iron contains 65.5% chlorine. Calculate the empirical formula.

**Answer**

| | Iron | Chlorine |
|---|---|---|
| $m/A_r$ | 34.5/55.8 | 65.5/35.5 |
| Moles | 0.618 | 1.85 |
| Ratio | 1 | 3 |

The ratio of iron:chlorine is 1:3, so the empirical formula is $FeCl_3$.

### Example 2

A compound has the empirical formula $CH_2O$. Its molar mass is $180\,g\,mol^{-1}$. Deduce the molecular formula.

**Answer**

formula mass of $CH_2O$ = 30.0

$M_r = 180.0$

$$\frac{180.0}{30.0} = 6.00$$

$\therefore$ molecular formula = 6 × empirical formula = $C_6H_{12}O_6$

# Balanced equations and associated calculations

You should be able to:

- write balanced equations for all the reactions studied
- balance equations for unfamiliar reactions where the reactants and products are specified

You can only gain marks for balancing an equation if all the formulae are correct. Therefore, you should make sure that you can work out the formulae of compounds.

## Reacting masses from balanced equations

The masses of the products and the reactants involved in a reaction can be deduced from the **stoichiometry** (the number of moles of each substance) of the balanced equation.

### Example 1

What is the mass of the solid residue when 10.0 g of potassium hydrogencarbonate are heated?

**Answer**

The equation for the thermal decomposition of potassium hydrogencarbonate is:

$$2KHCO_3(s) \rightarrow K_2CO_3(s) + CO_2(g) + H_2O(g)$$

2 mol    1 mol    1 mol    1 mol

$$\text{moles of } KHCO_3 \text{ used} = \frac{\text{mass}}{M_r}$$

$$= \frac{10.0}{39.1 + 1.0 + 12.0 + 48.0}$$

$$= \frac{10.0}{100.1} = 0.100$$

$$\therefore \text{ moles of } K_2CO_3 \text{ formed} = \frac{0.100}{2} = 0.0500$$

$$\text{mass of } K_2CO_3 \text{ formed} = \text{mol} \times M_r$$

$$= 0.0500 \times (78.2 + 12.0 + 48.0)$$

$$= 0.0500 \times 138.2 = 6.91 \text{ g}$$

### Example 2

What mass of quicklime (CaO) can be obtained by heating 75.0 g of limestone, which is 90.0% calcium carbonate?

**Answer**

The equation for the thermal decomposition of calcium carbonate is:

$$CaCO_3(s) \rightarrow CaO(s) + CO_2(g)$$

1 mol    1 mol    1 mol

$$\text{mass of calcium carbonate present in the limestone} = 75.0 \times \frac{90.0}{100} = 67.5 \text{ g}$$

$$\text{moles of CaCO}_3 \text{ decomposed} = \frac{\text{mass}}{M_r}$$

$$= \frac{67.5}{40.1 + 12.0 + 48.0}$$

$$= \frac{67.5}{100.1} = 0.674$$

$\therefore$ moles of CaO formed = 0.674

mass of CaO formed = mol × $M_r$

$= 0.674 \times (40.1 + 16.0)$

$= 0.674 \times 56.1 = 37.8\,\text{g}$

Some questions may appear to be more difficult than these examples. However, the steps in the calculations will be the same, so make sure that you familiarise yourself with these by doing lots of examples and always showing your working clearly.

Some questions that involve reacting masses are set on an industrial scale. This means that kilograms or tonnes may be used. Don't be intimidated by this type of question; if 1 g of substance X gives 2 g of substance Y, then 1 kg of X gives 2 kg of Y. The questions are basically the same, but with different units quoted in your answer.

## Calculating concentrations and volumes for reactions in solution

The type of calculations in the exam will vary, but the most straightforward ones can be divided into three main steps:

**Step 1** When you know both the concentration in mol dm$^{-3}$ and the volume ($V_1$), use the formula $\boldsymbol{n_1}$ **= concentration ×** $\boldsymbol{V_1}$**/1000** to calculate the number of moles of (solute in the) solution.

**Step 2** Write the equation for the reaction (or use the equation that has been given) and use the ratio of the reactants in the equation to predict the number of moles, $n_2$, of the other solution.

**Step 3** You know the values of $n_2$ (from step 2) and $V_2$ (different from the volume used in step 1), so rearranging $n_2$ = concentration × $V_2$/1000 allows you to determine the concentration of the other solution.

> ***Tip*** Remember to always quote your answer to the correct number of significant figures.

### Example 1

A 25.0 cm$^3$ portion of 0.100 mol dm$^{-3}$ sodium hydroxide required 22.5 cm$^3$ of ethanoic acid for complete reaction. Calculate the concentration of the ethanoic acid in mol dm$^{-3}$.

**Answer**

$$\text{number of moles of sodium hydroxide} = \text{concentration} \times \frac{V}{1000}$$

$$= 0.100 \times \frac{25.0}{1000} = 2.50 \times 10^{-3}\,\text{mol}$$

Using the equation, the number of moles of ethanoic acid can be deduced (1:1 ratio).

| $NaOH$ | + | $CH_3COOH$ | → | $CH_3COONa$ | + | $H_2O$ |
|---|---|---|---|---|---|---|
| 1 mol | | | | 1 mol | | |
| $2.50 \times 10^{-3}$ mol | | | | $2.50 \times 10^{-3}$ mol | | |

The concentration of ethanoic acid in mol dm$^{-3}$ is found using:

$$n = \text{concentration} \times \frac{V}{1000}$$

$$2.50 \times 10^{-3} = \text{concentration} \times \frac{22.5}{1000}$$

Rearranging gives: concentration $= 2.50 \times 10^{-3} \times \frac{1000}{22.5} = 0.111\,\text{mol dm}^{-3}$

**Example 2**

What volume of 0.185 mol dm$^{-3}$ sodium hydroxide solution reacts completely with 20.0 cm$^3$ of 0.100 mol dm$^{-3}$ sulfuric acid?

**Answer**

$$\text{number of moles of sulfuric acid} = \text{concentration} \times \frac{V}{1000}$$

$$= 0.100 \times \frac{20.0}{1000} = 2.00 \times 10^{-3}\,\text{mol}$$

Using the equation, the number of moles of sodium hydroxide can be deduced (1:2 ratio):

| $H_2SO_4(aq)$ | + | $2NaOH(aq)$ | → | $Na_2SO_4(aq)$ | + | $2H_2O(l)$ |
|---|---|---|---|---|---|---|
| 1 mol | | 2 mol | | | | |
| $2.00 \times 10^{-3}$ mol | | $4.00 \times 10^{-3}$ mol | | | | |

The concentration of sodium hydroxide in mol dm$^{-3}$ is found using:

$$n = \text{concentration} \times \frac{V}{1000}$$

Rearranging gives: $V = n \times \frac{1000}{\text{concentration}}$

$$= 4.00 \times 10^{-3} \times \frac{1000}{0.185}$$

$$= 21.6\,\text{cm}^3$$

## Calculating percentage yield

Calculations to find the mass of product using the balanced equation assume that the yield of the reaction is 100%. However, in practice, this is extremely rare. The percentage yield for a particular reaction can be calculated using the following relationship:

$$\%\text{ yield} = \frac{\text{mass of product obtained}}{\text{calculated (expected) mass}} \times 100$$

**Example**

Magnesium sulfate crystals, $MgSO_4.7H_2O$, can be obtained by allowing the solution formed by the reaction of magnesium with sulfuric acid to crystallise. When 1.80 g of magnesium was added to a slight excess of sulfuric acid and the solution allowed to crystallise, 10.83 g of magnesium sulfate crystals were obtained. Calculate the % yield in this experiment.

**Answer**

$$Mg + H_2SO_4 + 7H_2O \rightarrow MgSO_4.7H_2O + H_2$$

$$\text{moles of magnesium used} = \frac{\text{mass used}}{M_r}$$

$$= \frac{1.80}{24.3} = 0.0741$$

moles of magnesium sulfate formed = 0.0741

$M_r$ ($MgSO_4.7H_2O$) = 24.3 + 32.1 + (4 × 16.0) + (7 × 18.0) = 246.4

mass of crystals expected = moles × $M_r$ = 0.0741 × 246.4 = 18.25 g

$$\%\text{ yield} = \frac{10.83}{18.25} \times 100 = 59.336 = 59.3\%$$

## Calculating percentage atom economy

**Atom economy** is defined as the percentage of the atoms of the reactants that are converted into useful products and are, therefore, not wasted. For example, the production of hydrogen chloride by the direct reaction of hydrogen with chlorine has a 100% atom economy. The percentage atom economy for a particular product can be calculated from balanced equations using the following relationship:

$$\%\text{ atom economy} = \frac{\text{mass of desired product}}{\text{total mass of reactants}} \times 100$$

**Example**

Quicklime, CaO, is produced by the thermal decomposition of limestone, $CaCO_3$, according to the equation:

$$CaCO_3(s) \rightarrow CaO(s) + CO_2(g)$$

Calculate the percentage atom economy of the process.

**Answer**

$$M_r\ (CaCO_3) = 40.1 + 12.0 + (3 \times 16.0) = 100.1$$

$$M_r\ (CaO) = 40.1 + 16.0 = 56.1$$

Therefore, theoretically 100.1 g limestone could produce 56.1 g quicklime.

$$\%\text{ atom economy} = \frac{\text{mass of desired product}}{\text{total mass of reactants}} \times 100$$

$$= \frac{56.1}{100.1} \times 100 = 56.0\%$$

In an exam question, the number of reactants and/or the masses involved may not be as simple as in the above example. You should make sure that you understand the principle involved and how to apply it.

***Tip*** Do not confuse percentage yield and percentage atom economy.

# Bonding

## The nature of ionic, covalent and metallic bonds

Try to remember the characteristic features of each type of bonding as a list and then convert the list into the sentences that you need for the exam. Your answers should be concise and include only relevant material. You must be able to recognise the type of bonding when presented with different elements or compounds.

### Ionic bonding

The main points are:

- transfer of electrons
- transfer is from metal atoms to non-metal atoms
- metal atoms lose $e^-$ to form cations (positive ions)
- non-metal atoms accept $e^-$ to form anions (negative ions)
- oppositely charged ions are held together by strong electrostatic forces of attraction
- the forces of attraction result in a **giant**, three-dimensional, ionic, crystalline **lattice**

Examples of ionically bonded compounds include sodium chloride (NaCl), potassium bromide (KBr) and magnesium oxide (MgO).

The ionic lattice of sodium chloride is illustrated on page 36.

## Covalent bonding

The main points are:

- electrons are shared
- sharing is between two non-metals
- non-metal atoms both need to gain electrons — this is achieved by sharing
- a **covalent bond** is defined as a **shared pair of electrons**
- covalent bonds can exist in isolation in simple molecules

Examples of covalently bonded substances include chlorine ($Cl_2$), fluorine ($F_2$), methane ($CH_4$), ammonia ($NH_3$), water ($H_2O$), hydrogen chloride (HCl) and hydrogen fluoride (HF).

A simple molecule of hydrogen fluoride is shown below:

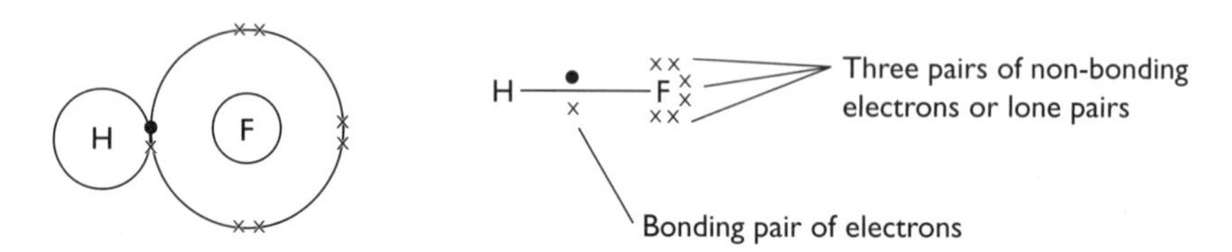


## Coordinate or dative covalent bonding

The main points are:

- A covalent bond is defined as a shared pair of electrons.
- A coordinate bond or a dative covalent bond is defined as a *shared pair of electrons* where *both electrons have originated from one atom.*
- Once a coordinate bond has formed, it is identical to a covalent bond because it is a shared electron pair.

A coordinate bond is formed when ammonia gas and boron trifluoride gas react to give a white solid with the composition $NH_3BF_3$:

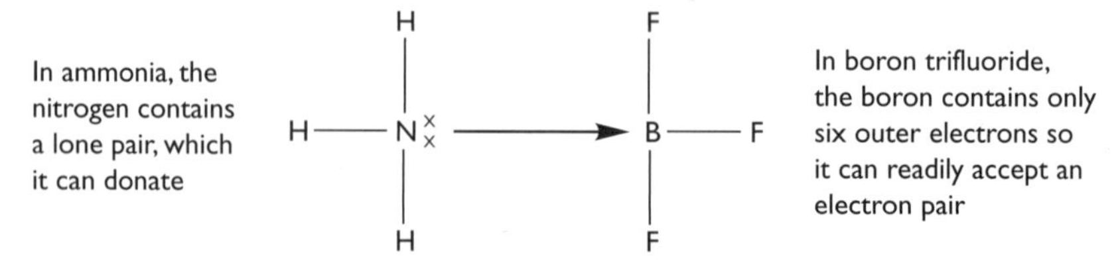


A further example of a coordinate bond is when ammonia reacts with a hydrogen ion to produce an ammonium ion:

$H_3N\!:\rightarrow H^+$ produces $NH_4^+$

### Metallic bonding

The main points are:

- A metal is a giant three-dimensional lattice, consisting of closely packed metal ions surrounded by delocalised electrons.
- Delocalised electrons are free to move through the lattice.
- A metallic bond is the electrostatic force of attraction that two neighbouring cations have for the delocalised electrons between them.

Examples include any metal, such as sodium, potassium, calcium, magnesium and iron. A lattice of magnesium metal is illustrated on page 37.

# Bond polarity

You will encounter questions on bond polarity if the covalent molecule contains atoms with differing electronegativities. You need to understand the concept of electronegativity.

An electron pair shared equally between two atoms constitutes a covalent bond. Sometimes, this electron pair is not shared equally because some atoms attract electrons more than others. The unequal sharing of electrons is known as **bond polarity**. It represents an introduction of some ionic character into the covalent bond.

**Electronegativity** is a measure of the relative tendency of an atom to attract a bonding pair of electrons (or to withdraw electron density) from a covalent bond. Elements have electronegativity values ranging from 0 to 4. You need to remember the trends, not the actual numbers.

Important points about electronegativity include:

- Electronegativity increases across a period.
- Electronegativity decreases down a group.
- Fluorine is the most electronegative element in the periodic table.
- Small atoms with a large number of protons in the nucleus attract electron density more strongly.
- No values are quoted for the noble gases because these elements generally do not form compounds with covalent bonds.

When a covalent bond exists between atoms of differing electronegativity, the shared pair of electrons is displaced towards the more electronegative atom. The charge separation creates an electric dipole and the bond is described as polar. The greater the difference in electronegativity between the elements, the more polar is the bond.

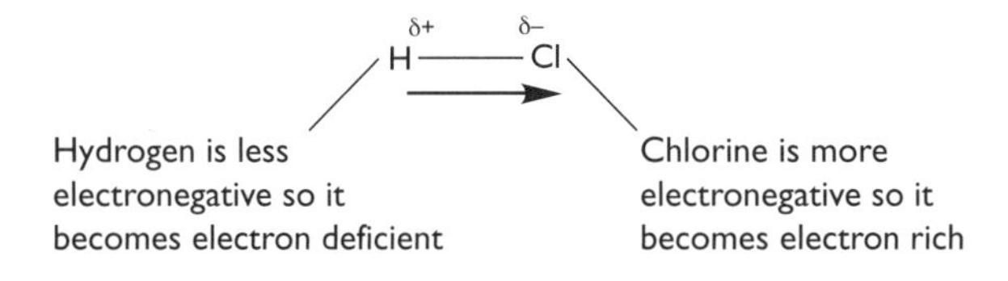

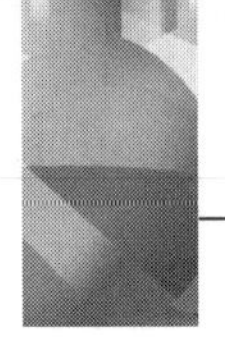



In **symmetrical molecules**, such as carbon dioxide, the bond polarities cancel, so there is *no* overall polarity.

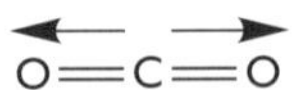


In **asymmetrical molecules**, such as water, the bond polarities do not cancel. There is an overall polarity and the molecule is polar.

# Forces acting between molecules

You need to understand the different types of intermolecular force so that you can explain the physical properties of substances, such as boiling points and solubility.

There are three types of intermolecular force. In order of increasing strength they are:

- induced dipole–dipole (van der Waals) forces
- permanent dipole–dipole
- hydrogen bonding

## Predicting the type of intermolecular force

**Induced dipole–dipole** (van der Waals) forces attract all covalent molecules to each other.

Polar molecules contain atoms with different electronegativities, so in addition to induced dipole–dipole (van der Waals) forces, the molecules also attract each other by **permanent dipole–dipole** forces.

If a molecule contains hydrogen, which is directly bonded to a nitrogen, oxygen or fluorine atom, it sets up an extreme form of permanent dipole–dipole force between the molecules. This is **hydrogen bonding**.

### An example of induced dipole–dipole (van der Waals) forces: iodine

- Iodine ($I_2$) is a non-polar molecule.
- The iodine atoms have identical electronegativities.
- Electron distribution is usually symmetrical.
- A temporary dipole occurs when electron distribution is asymmetric (unequal) due to the fluctuating movement of electrons.
- Iodine molecules are large and contain a large number of electrons. The induced dipole–dipole forces between the molecules are therefore relatively strong compared with those in other simple, non-polar molecules that have fewer electrons.

Remember, compared with other intermolecular forces, the induced dipole–dipole forces in iodine are weak. However, they are strong enough for iodine molecules to form solid crystals at room temperature.

All covalent molecules have induced dipole–dipole forces between them — for example, chlorine, bromine, methane, phosphorus and sulfur.

### An example of permanent dipole–dipole forces: hydrogen chloride

- Hydrogen chloride (HCl) is a polar molecule.
- The hydrogen and chlorine atoms have different electronegativities.
- Chlorine has a greater electronegativity than hydrogen.
- Electron distribution is asymmetrical (i.e. permanent charge separation).
- Permanent dipole–dipole forces exist between the molecules.
- When comparing molecules of similar size, permanent dipole–dipole forces are stronger than induced dipole–dipole (van der Waals) forces.

Examples of other molecules with permanent dipole–dipole forces include chloromethane and hydrogen sulfide.

### An example of hydrogen bonding: water

- Water ($H_2O$) is a polar molecule.
- The hydrogen is directly attached to a small, very electronegative, oxygen atom.
- Oxygen is electron-rich and hydrogen is electron-deficient.
- The electron-deficient hydrogen atom attracts a lone pair from an oxygen atom on a neighbouring water molecule.
- When comparing molecules of similar size, hydrogen bonds are stronger than permanent dipole–dipole forces.

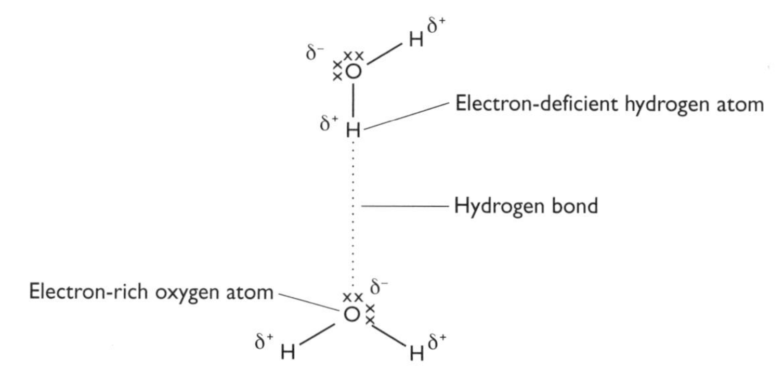


Other examples include ammonia, hydrogen fluoride and methanol.

# States of matter

You need to be able to explain the energy changes associated with changing states.

Therefore, you need to have some understanding of the movement and positions of particles in solids, liquids and gases. and of the forces acting between the particles.

In **solids**:

- there is a high degree of order
- particles vibrate about a fixed position
- the forces holding a solid together vary, but are usually stronger than those in the liquid and gas states of the same substance

In a solid, an increase in temperature gives the particles more energy and they vibrate faster. At the melting point, the temperature remains constant because the energy is used to break the forces of attraction between the particles. This energy is called the **enthalpy of fusion**.

In **liquids**:

- the order is intermediate between a solid and a gas, but is closer to that of a solid
- the separation of particles in liquids is only slightly greater than in solids
- forces are still present between the particles, but are weaker than in the solid, so particles can move relative to each other throughout the bulk of the liquid

In a liquid, if the temperature increases, the kinetic energy of the particles in the liquid increases until the boiling point. The temperature remains constant when the remaining forces of attraction are broken sufficiently for the particles to move apart from each other. The energy supplied is called the **enthalpy of vaporisation**.

In **gases**:

- there is a high degree of disorder
- the particles move with rapid, random motion
- the forces of attraction between the particles are *negligible*, particularly at low pressures and high temperatures when the particles are far apart

## Types of crystal

You are expected to be able to draw simple sketches (not of textbook standard) of the crystal structures. You should also be able to state and explain the physical properties of the crystals in terms of structure and bonding. When discussing physical properties, remember three key points: **melting points**, **solubility** and **conductivity**.

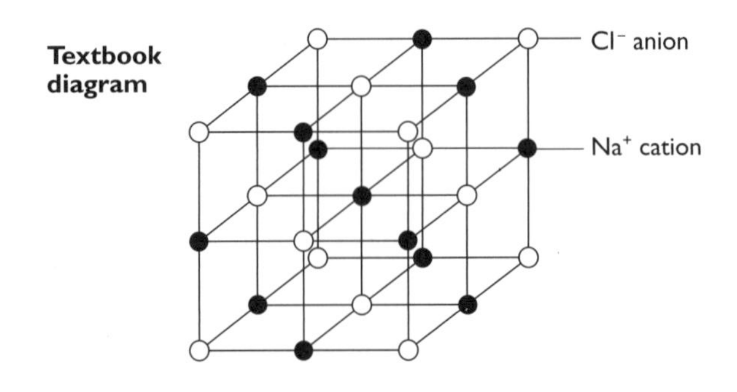


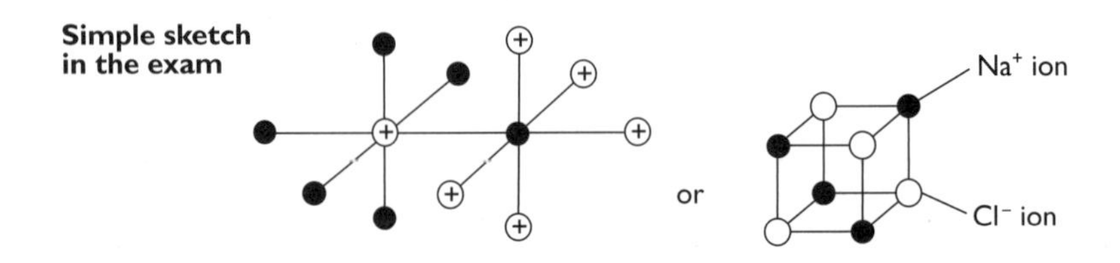

Crystals are solids with particles organised in a regular structure. They are classified according to the type of bonding between the particles. There are four types:

- ionic
- metallic
- molecular
- macromolecular

## Ionic crystals

Example: sodium chloride, NaCl

### *Structure*

Sodium chloride is a giant, ionic, three-dimensional lattice of oppositely charged ions with strong electrostatic forces between these ions.

### *Properties*

- High melting point, because of strong electrostatic forces between ions.
- Soluble in polar solvents, because the ions interact with polar water molecules and the lattice breaks down.
- Insulator in the solid state, because the ions can only vibrate about a fixed position.
- Conducts when molten or when in solution, because the ions are mobile.

## Metallic crystals

Example: magnesium, Mg

### *Structure*

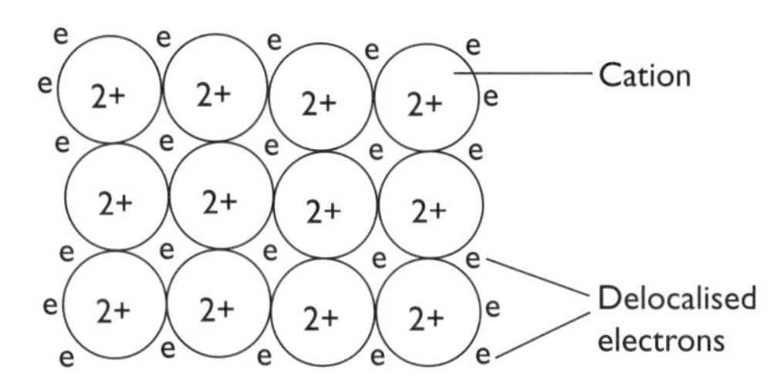


Metallic crystals consist of a three-dimensional metallic lattice of closely packed cations in a sea of delocalised electrons. There are strong forces between the cations and the delocalised electrons.

> ***Tip*** The typical textbook diagram and the simple sketch in the exam are the same. Make sure that you show the cations in a **close-packed arrangement**. The total number of delocalised electrons should equal the total charge of the cations.

### *Properties*

- Relatively high melting point, because of the strong metallic bonds.
- Conducts when solid, due to delocalised electrons which can move through the lattice.
- Insoluble in water, but may react with water.
- Malleable and ductile because the layers of ions can slide over each other.

## Molecular crystals

Molecular crystals contain covalent molecules held together by induced dipole–dipole forces, permanent dipole–dipole forces, or hydrogen bonding.

Examples: iodine, $I_2$, and ice, $H_2O$

### *Structure of iodine*

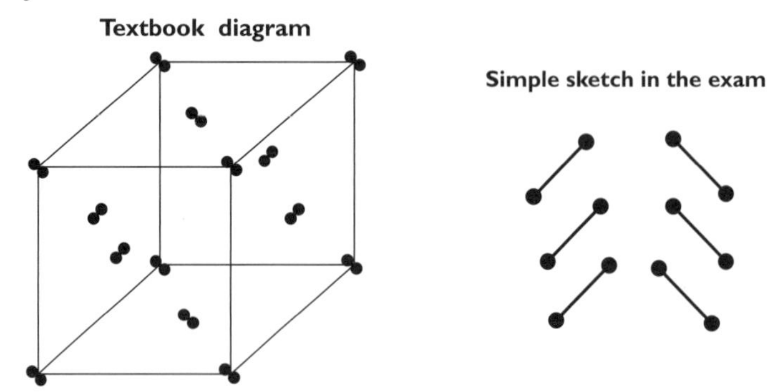


Iodine has a molecular lattice with weak, temporary, induced dipole–dipole (van der Waals) forces between the molecules. There are strong covalent bonds between the iodine atoms within each molecule.

### *Properties of iodine*

- Iodine sublimes because the weak forces *between* molecules are easily overcome
- It is almost insoluble in water but dissolves in organic solvents.
- It does not conduct because the electrons are localised and not free to move.

### *Ice, $H_2O(s)$*

In liquid water, the molecules are associated by hydrogen bonds in varying, continuously changing combinations. When water freezes, the molecules arrange themselves so that each oxygen atom forms two hydrogen bonds with hydrogen atoms of other water molecules. This results in fixed, three-dimensional, tetrahedral arrangements of hydrogen bonds and, in the resulting lattice structure, six water molecules form hexagonal rings. This lattice structure holds the water molecules further apart than in the liquid, which results in ice being less dense than water.

## Macromolecular (giant atomic) crystals

Examples: diamond and graphite

### *Structure of diamond*

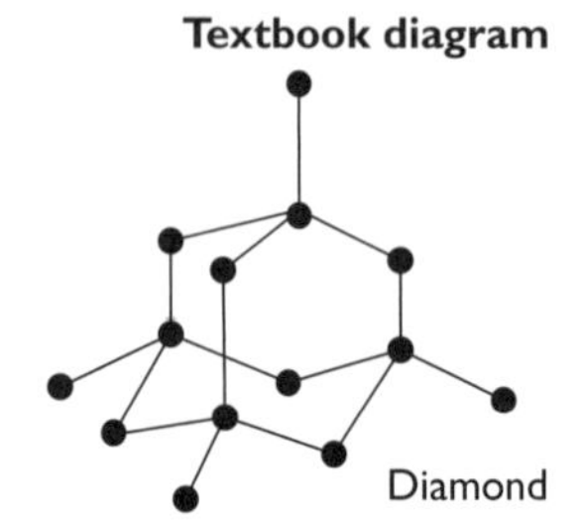


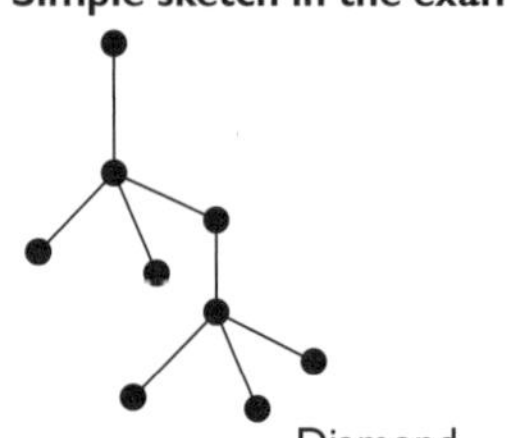

- The structure of diamond is a three-dimensional tetrahedral lattice.
- Each carbon atom is joined to four others by strong covalent bonds.

*Note*: Silicon has a similar structure to diamond.

***Structure of graphite***

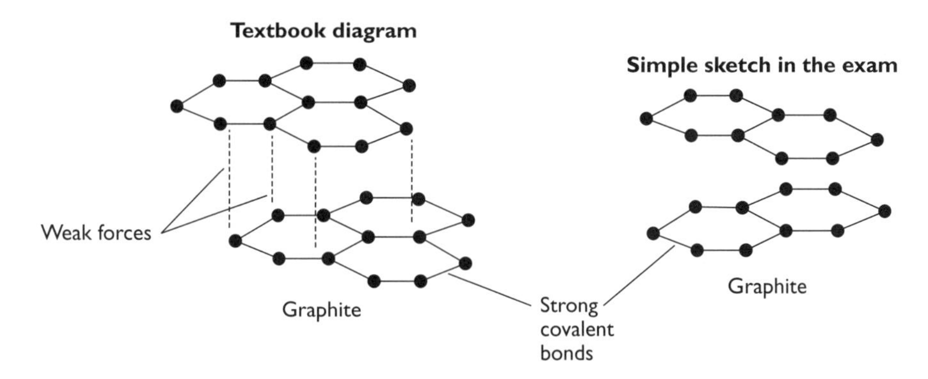


- Graphite consists of layers of carbon atoms, arranged hexagonally.
- Each carbon atom is covalently bonded to three others within the layers and weak forces hold the layers together.

***Properties of diamond and graphite***

- Both diamond and graphite have extremely high melting points because the strong covalent bonds have to be broken before melting can occur.
- Diamond does *not* conduct electricity because the electrons are localised in the bonds.
- Graphite conducts electricity because the lattice contains mobile delocalised electrons between the layers.

# Shapes of simple molecules and ions

## Basic principles

- Shape depends on the number of electron pairs around the central atom.
- Electron pairs maximise their distance apart in order to minimise repulsion.
- There are bonding pairs (bpr) and lone pairs (lpr).
- The order of repulsion is **lpr/lpr > lpr/bpr > bpr/bpr**.

## Basic shapes

When bonding pairs only are involved, there are five basic shapes to remember:

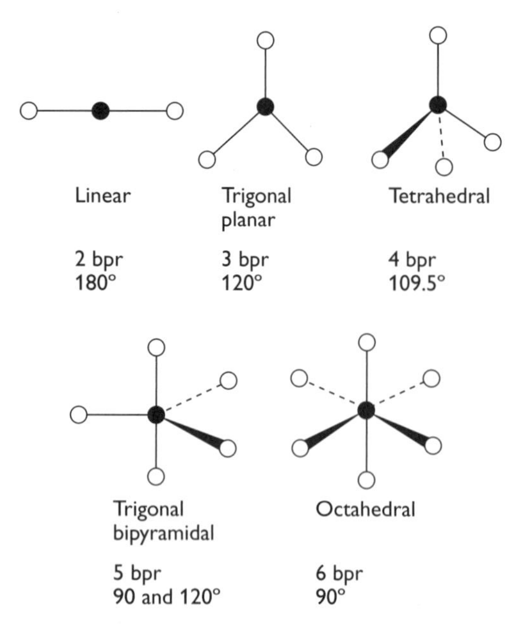


***Tip*** When drawing any sketch in the exam, make sure that you include the symbols of the atoms involved, not just the circles. For example, the shape of a water molecule should be shown as not

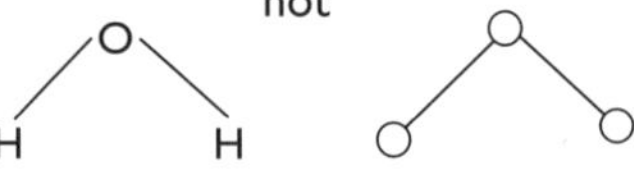


## The effect of lone pairs

The order of repulsion is lpr/lpr > lpr/bpr > bpr/bpr, so the presence of a lone pair reduces the angle between the bonded pairs. For example, in molecules with four electron pairs:

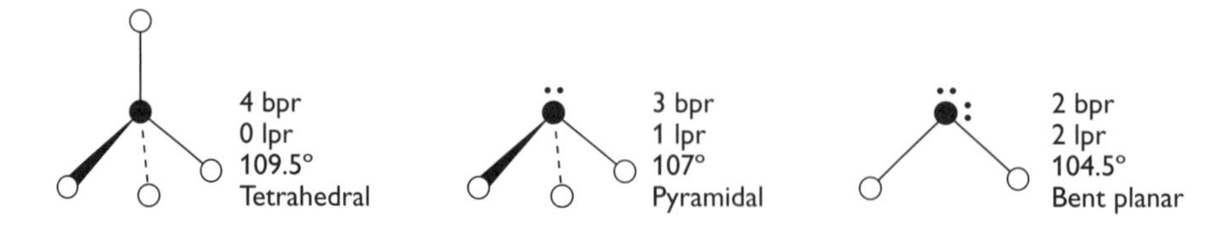


An easy way to predict the bond angles is to start with the tetrahedral shape and then take 2.5° off for every lone pair present around the central atom.

## A method for determining shapes of molecules

You are required to be able to predict the shapes of molecules and ions that contain only single covalent bonds. The following method works for all the molecules and ions on the AQA specification.

| Method | Examples | | | | |
|---|---|---|---|---|---|
| | $BF_3$ | $SF_6$ | $PCl_3$ | $PCl_4^+$ | $PCl_6^-$ |
| Count the total number of electrons in the outer shell of the central atom | 3 | 6 | 5 | 5 | 5 |
| Add 1 if negative ion; subtract 1 if positive ion | 0 | 0 | 0 | −1 | +1 |
| Add 1 for each bonded atom | 3 | 6 | 3 | 4 | 6 |
| Total the number of electrons | 6 | 12 | 8 | 8 | 12 |
| Divide by 2 to obtain the number of electron pairs | 3 | 6 | 4 | 4 | 6 |
| Count the number of bonded atoms — the extra pairs are lone pairs | 3 bpr | 6 bpr | 3 bpr, 1 lpr | 4 bpr | 6 bpr |
| Shape | Trigonal planar | Octahedral | Trigonal pyramidal | Tetrahedral | Octahedral |
| Bond angle | 120° | 90° | 107° | 109.5° | 90° |

## Predicting the shapes of more complex molecules

Do not panic when you are faced with questions of this type; just apply the basic principles set out above. Let us try it using the examples of xenon tetrafluoride ($XeF_4$) and sulfur tetrafluoride ($SF_4$).

### $XeF_4$

Xenon (Xe) has eight outer electrons. Add another four for the bonded fluorine atoms and then divide by two. This gives six electron pairs; so the molecule is based on an octahedral shape. There are four bonding pairs and two lone pairs, so the shape is **square planar**. The lone pairs are the maximum distance apart.

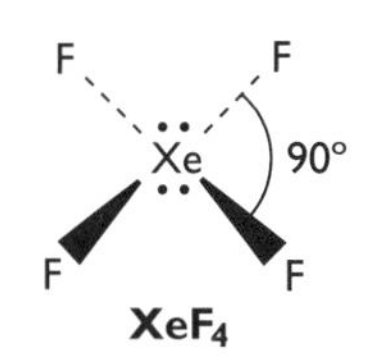


$XeF_4$

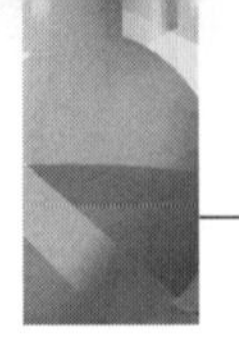

**$SF_4$**

Sulfur (S) has six outer electrons. Add another four for the bonded fluorine atoms and divide by two. This gives five electron pairs; so the molecule is based on a **trigonal bipyramidal** shape. However, the lone pair distorts the usual bond angles.

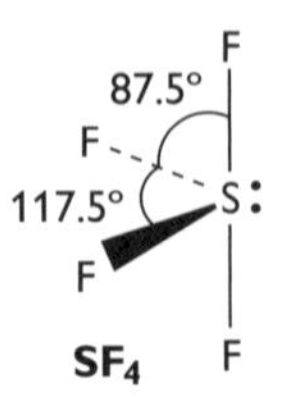


# Periodicity

## Classification of elements in *s*, *p* and *d* blocks

You need to be able to classify an element as *s*, *p* or *d* block, according to its position in the periodic table.

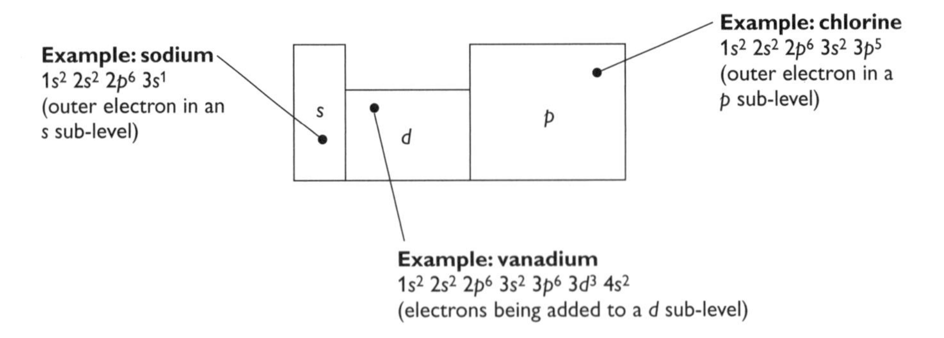


## Properties of the elements of period 3

You need to be able to state and explain, in terms of structure and bonding, the trends in atomic radius, first ionisation energy, and melting and boiling points of the elements sodium to argon.

The examiner will ask you to state the trend, for 1 mark, and then to explain it, usually for an additional 3 marks. If your statement of the trend is incorrect, you will lose all the marks for the explanation, so it is essential that you know and understand each of the trends.

## The trend in atomic radius and ionisation energy

You have to be able either to make a statement or to draw a graph describing the trend in the properties, as shown below:

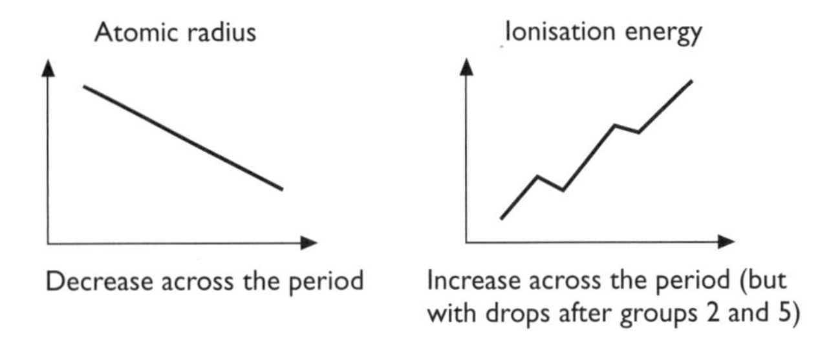


### Explanation of trend

You can use the same answer for both the properties shown above, except for a slight change in the last sentence.

***Atomic radius***

- The number of protons (or nuclear charge) increases.
- The shielding stays the same (or electrons enter the same energy level).
- The attraction of the nuclear charge for the electrons is greater, so *the atoms shrink.*

***Ionisation energy***

- The number of protons (or nuclear charge) increases.
- The shielding stays the same (or electrons enter the same energy level).
- The attraction for the electrons is greater, so *the outer electron is harder to lose.*

## The trend in melting point

Questions on this topic occur frequently. You should to be able to describe in detail the trend in melting point across period 3 and then relate this trend to the structure and bonding found in the elements.

The trend can be divided into three main sections, as shown on the graph below:

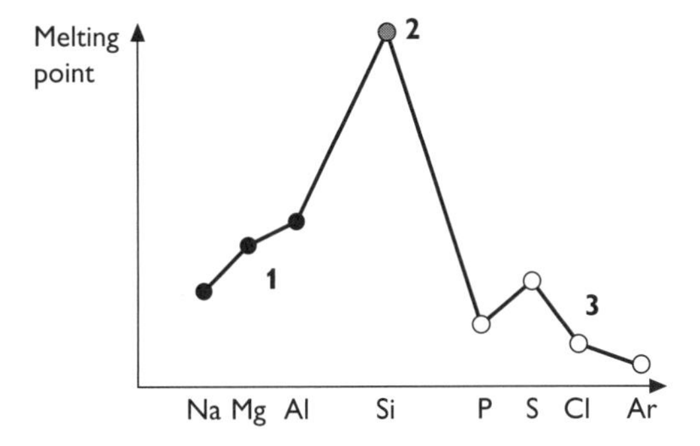


**Section 1** of the graph:

- The elements sodium (Na), magnesium (Mg) and aluminium (Al) are all metallic.

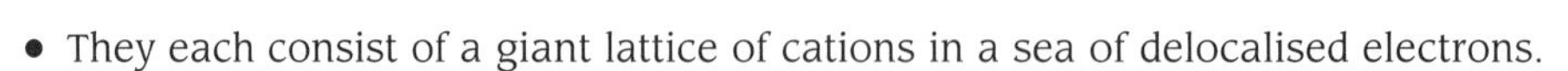

- They each consist of a giant lattice of cations in a sea of delocalised electrons.
- Melting point increases because metallic bond strength increases.
- Bond strength increases because the charge of the cations increases; the cations are smaller and there are more electrons per cation, which all contribute to an increased attraction.

**Section 2** of the graph:
- Silicon (Si) has an extremely high melting point.
- It is macromolecular.
- Silicon atoms are arranged in a tetrahedral fashion, in a giant lattice similar to that of diamond.
- Strong covalent bonds link all the silicon atoms together.

**Section 3** of the graph:
- Phosphorus ($P_4$), sulfur ($S_8$) and chlorine ($Cl_2$) are simple covalent molecules.
- Their melting points are relatively low because only weak induced dipole–dipole (van der Waals) forces exist between the molecules.
- These forces increase with the total number of electrons or molecular size.
- Sulfur ($S_8$) has the highest melting point of the elements in this section and argon has the lowest.
- Argon (Ar) exists as individual atoms, which are not very polarisable, so the induced dipole–dipole forces are very weak.

# Introduction to organic chemistry

## Definitions

- The **empirical formula** is the simplest whole number ratio of atoms of each element in a compound.
- The **molecular formula** is the actual number of atoms of each element in a compound.
- The **structural formula** shows the arrangement of the atoms and bonds within the molecule.
- A **functional group** is an atom or group of atoms that gives the organic compound its characteristic chemical properties.
- A **homologous series** is a family of organic compounds that possess the same general formula, show similar chemical properties and show a gradual change in physical properties. For example, the general formula of the alkanes is $C_nH_{2n+2}$; the general formula of the alkenes is $C_nH_{2n}$.

# Naming organic compounds

Organic compounds are named according to the following rules:

- The root of the name is derived from the number of carbon atoms present in the longest **unbranched** chain:

| Number of carbon atoms | 1 | 2 | 3 | 4 | 5 | 6 |
|---|---|---|---|---|---|---|
| Root | Meth- | Eth- | Prop- | But- | Pent- | Hex- |

- If the carbon chain is joined up as a **ring**, then the prefix **cyclo**- is used.
- A chain can be **saturated** or **unsaturated**. This is indicated by the second syllable. If the chain is saturated (single carbon-to-carbon bonds), the second syllable is '**-an-**' — for example ethane, $CH_3CH_3$. If the chain is unsaturated (double carbon-to-carbon bond), the second syllable is '**-en-**' — for example ethene, $CH_2{=}CH_2$.
- A prefix or suffix is used to indicate the presence of a functional group. The prefix **chloro-**, **bromo-** or **iodo-** indicates a haloalkane. The suffix **-ol** indicates an alcohol, for example ethanol, $CH_3CH_2OH$. If no other functional group requiring a suffix is present, then the name ends in the letter **e**.

  | $CH_3CH_3$ | $CH_2{=}CH_2$ | $CH_3CH_2OH$ |
  |---|---|---|
  | Ethane | Ethene | Ethanol |

- In many organic molecules, the carbon skeleton is **branched**. The names of the **side chains** depend on the number of carbon atoms present: $CH_3$ is methyl; $CH_3CH_2$ is ethyl; $CH_3CH_2CH_2$ is propyl.

$$\begin{array}{c} \quad\;\; CH_3 \\ \quad\;\; | \\ CH_3-CH-CH_2-CH_3 \end{array}$$

Methylbutane

- A *number* indicates the *position* of the functional group or side chain on the main carbon skeleton. For example:

  | $CH_3CH_2CH_2Cl$ | $CH_3CHICH_3$ |
  |---|---|
  | 1-chloropropane | 2-iodopropane |

- Carbon atoms are numbered consecutively from one end, such that the attached groups are on the *lowest numbered* carbon atoms. For example:

$$\begin{array}{c} 4 \qquad\; 3 \qquad\;\; 2 \qquad 1 \\ CH_3-CH_2-CH-CH_3 \\ \qquad\qquad\quad | \\ \qquad\qquad\quad Br \end{array}$$

2-bromobutane

- To indicate the *position of a double bond*, the number of the lowest numbered carbon atom involved is placed before –ene. For example, $CH_2{=}CHCH_2CH_3$ is called but-1-ene; $CH_3CH{=}CHCH_3$ is but-2-ene.
- If the same functional group appears more than once, this is indicated by a prefix. **Di-** means *two* groups, **tri-** means *three* groups and **tetra-** means *four* groups.

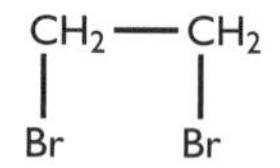


1,2-dibromoethane

- If more than one type of functional group is present, then positions and names are listed in alphabetical order of the functional group, i.e. bromo- before chloro- and tri**b**romo- before di**c**hloro-.

$CH_2$—CH—$CH_3$ (Cl on $C_1$, Br on $C_2$)

2-bromo-1-chloropropane

# Isomerism

Isomerism occurs where molecules with the same molecular formula have a different arrangement of atoms. There are two main types of isomerism:

- structural isomerism
- stereoisomerism

Both types of isomerism can be further subdivided.

*Note*: stereoisomerism is covered in Unit 2.

## Structural isomerism

**Structural isomerism** occurs when there are two or more compounds with the same molecular formula but with a different structural formula.

**Chain isomerism** and **position isomerism** are different types of structural isomerism.

### Chain isomerism

- Chain isomerism occurs when there are two or more ways of arranging the **carbon skeleton**.
- Chain isomers have similar chemical properties but slightly different physical properties.
- The more branched the isomer, the weaker are the induced dipole–dipole (van der Waals) forces and the lower the boiling point.
- The isomers of $C_5H_{12}$ are examples of chain isomerism:

$CH_3$—$CH_2$—$CH_2$—$CH_2$—$CH_3$

Pentane

$CH_3$—CH($CH_3$)—$CH_2$—$CH_3$

Methylbutane

$CH_3$—C($CH_3$)($CH_3$)—$CH_3$

Dimethylpropane
(lowest boiling point)

**Position isomerism**

- Position isomers have the same carbon skeleton and the same functional group.
- The functional group is joined at *different* places on the carbon skeleton.
- The isomers of $C_3H_7Br$ are examples of position isomerism:

$CH_3—CH_2—CH_2—Br$
1-bromopropane

$CH_3—CH(Br)—CH_3$
2-bromopropane

# Alkanes

You need to know the components of petroleum (crude oil) and how the different fractions are separated.

You need to understand the principle of cracking and know that different types of cracking are used to produce different products for the petrochemical industry.

The chemistry of alkanes is limited because of their unreactive nature. The emphasis in this section is on their combustion.

# Crude oil and fractional distillation

Crude oil is:

- a dark, viscous liquid
- a mixture of many different hydrocarbons (mainly alkanes) with a trace of sulfur compounds
- formed by decomposition, under intense heat and pressure, of marine and plant life over millions of years

## The separation of crude oil

The method used to separate crude oil is **fractional distillation**:

- Crude oil is separated into a number of simpler mixtures (**fractions**), which are useful as fuels, lubricants and as a source of petrochemicals.
- The fractions contain different molecules that have **different boiling point ranges**.
- The crude oil is vaporised and passed into a tall fractionating column.
- A **temperature gradient** exists in the column, which is hot at the bottom and cooler at the top.
- The fractions cool and condense at different levels — the smaller the molecule, the weaker are the forces between the molecules, the greater is its volatility and the higher in the column it condenses.
- The crude-oil residue is distilled further under vacuum. Lowering the pressure reduces the boiling point and ensures the constituents distil off below their decomposition temperature (i.e. it avoids cracking).

***Tip*** **You need to understand the principle that small molecules with weak intermolecular forces are found at the top of the fractionating column, where it is cooler.**

The demand for the fractions is not uniform. Demand for fractions containing alkanes of low $M_r$ exceeds supply. The supply of fractions containing alkanes of high $M_r$ exceeds demand. Alkanes of high $M_r$ are therefore broken down (**cracked**) to produce smaller-chain alkanes (more useful as fuels) and alkenes (which can be used to produce polymers and other chemicals).

# The cracking of alkanes

Cracking is the breaking of large alkanes into smaller molecules:

$$\text{Large } M_r \text{ alkane} \xrightarrow{\text{heat}} \text{Smaller } M_r \text{ alkane} + \text{alkene (or hydrogen)}$$

- Cracking can resolve the imbalance in the supply and demand of the fractions.
- Many of the smaller molecules are more useful as fuels such as petrol.
- Small, unsaturated molecules such as ethene and propene are also produced. These alkenes are used to make polymers and other chemicals.
- High temperatures required for cracking cause C–C (and C–H) bonds in alkanes to break, producing smaller alkane molecules and alkenes.
- In addition, other processes take place, such as **dehydrogenation** (loss of hydrogen to convert alkanes to alkenes and cycloalkanes to aromatic compounds), **cyclisation** (converting straight chains to cyclic compounds) and **isomerisation** (converting straight-chain alkanes to branched alkanes).
- The long-chain molecules can break up in a number of different ways to produce a mixture of products, which can be separated by fractional distillation.

You may be asked to construct a number of different equations to describe the process of cracking. The examiner will give you all the information you need in the question. Remember that the original, large molecule cracked will be an alkane, which breaks down to give a shorter-chain alkane and an unsaturated alkene. Always remember to check that the numbers of carbon atoms and hydrogen atoms on each side of the equation are exactly the same.

The following equations show the cracking of a long-chain alkane with 14 carbon atoms:

$$C_{14}H_{30} \rightarrow C_7H_{16} + C_3H_6 + 2C_2H_4$$

$$C_{14}H_{30} \rightarrow C_{12}H_{24} + C_2H_4 + H_2$$

$$C_{14}H_{30} \rightarrow C_8H_{18} + C_6H_{12}$$

The equations you write will depend on the information given by the examiner.

## Types of cracking

The type of cracking and the proportions of final products from the cracking process depend on the conditions employed.

### Thermal cracking

Important points include:

- High temperatures (800–900°C) and high pressures are used.
- A high proportion of alkenes is produced.
- The higher temperature causes the chains to break near their ends, producing small alkenes such as ethene.

### Catalytic cracking

Important points include:

- Lower temperatures (450°C) and only a slight pressure are used.
- **Zeolite catalysts** — crystalline **aluminosilicates** — are required.
- Branched alkanes, cycloalkanes and aromatic hydrocarbons are produced.
- Branched alkanes burn more uniformly and so are more useful as motor fuels.

# The combustion of hydrocarbons

Hydrocarbons produce a variety of products on combustion.

Limited (incomplete) combustion of hydrocarbons produces carbon or carbon monoxide and water. Complete combustion of hydrocarbons produces carbon dioxide and water. The three equations for the combustion of methane are:

$$CH_4 + O_2 \rightarrow C + 2H_2O$$

$$CH_4 + 1\tfrac{1}{2}O_2 \rightarrow CO + 2H_2O$$

$$CH_4 + 2O_2 \rightarrow CO_2 + 2H_2O$$

The examiner could ask you to balance equations for the combustion of any hydrocarbon, a popular choice being equations for hydrocarbons such as $C_8H_{18}$.

## Pollutants and the internal combustion engine

- Incomplete combustion of $C_8H_{18}$ produces carbon monoxide:

  $C_8H_{18} + 8\tfrac{1}{2}O_2 \rightarrow 8CO + 9H_2O$

- The reaction of $N_2$ and $O_2$ (present in air) via the spark in the engine produces **nitrogen monoxide**:

  $N_2 + O_2 \rightarrow 2NO$

*Note*: Other oxides of nitrogen, $NO_x$, are also formed.

These pollutants cause problems because:

- carbon monoxide is **toxic**

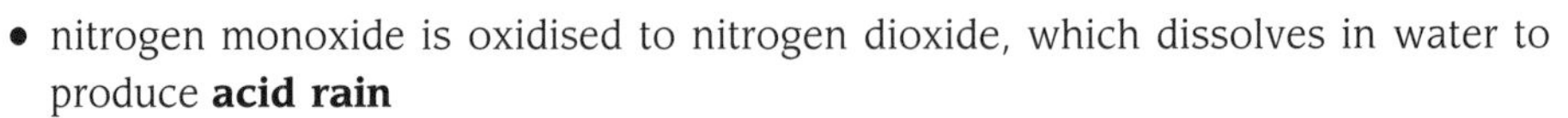

- nitrogen monoxide is oxidised to nitrogen dioxide, which dissolves in water to produce **acid rain**
- oxides of nitrogen ($NO_x$) react with unburnt hydrocarbons (also present if combustion is incomplete) to produce **photochemical smog**

**Removal of the pollutants by catalytic converters**

- The gases are passed through a ceramic honeycomb structure coated in precious metals, such as platinum, rhodium and palladium.
- The ceramic support provides the maximum surface area, with minimal use of the metal to save costs.
- Carbon monoxide and nitrogen monoxide react to produce nitrogen and carbon dioxide, which are less harmful products:

$$CO + NO \rightarrow CO_2 + \tfrac{1}{2}N_2$$

- The unburnt hydrocarbons react with nitrogen monoxide to produce nitrogen, carbon dioxide and water:

$$C_8H_{18} + 25NO \rightarrow 8CO_2 + 9H_2O + 12\tfrac{1}{2}N_2$$

## Other pollution arising from the combustion of petroleum fractions

Crude oil contains traces of sulfur compounds. When fractions containing these compounds are burned, sulfur dioxide is produced. This is a toxic gas that can cause breathing problems. It also reacts with water and oxygen in the atmosphere to produce sulfuric acid and, hence, acid rain. Today, most of the sulfur is removed from petrol for use in motor vehicles and the sulfur dioxide produced by the combustion of petroleum fractions in power stations and factories is removed by passing the 'flue gases' through beds (or 'scrubbers') containing calcium oxide. This is a base that reacts with the sulfur dioxide to form a salt.

## Greenhouse gases and global warming

Electromagnetic radiation from the sun is absorbed by the Earth's surface and is re-emitted into the atmosphere as infrared radiation. Much of this radiation is absorbed by molecules of gases such as carbon dioxide, methane and water vapour — collectively known as 'greenhouse gases' — which in turn re-emit a large proportion back to Earth. This is one of the processes that has helped to maintain the Earth at a temperature suitable for life. The percentage of carbon dioxide in the atmosphere had remained constant at about 0.03%, but since the beginning of the twentieth century this percentage has steadily increased, largely due to the increased use of fossil fuels. This has resulted in an increase in global temperatures.

# Questions & Answers

These questions are similar in style and content to those that you can expect in Unit Test 1. Although the number of questions is limited, they have been designed to test most of the key facts and concepts covered in Unit 1.

Unit Test 1 is divided into sections A and B. Section A questions are structured, with spaces at the end of each question part for your response. The number of questions in Section A varies, as do the mark allocations. The total mark for Section A is 50.

Section B contains one longer question, worth 20 marks, which is also divided into sections. It is effectively a long structured question, without answer spaces. The part questions are worth more marks than those in Section A.

All the questions in this Questions and Answers section are structured, without answer spaces. You should use the number of marks as a guide to the length of the answer required. If a question is worth 4 marks, then you should expect to write no more than four sentences, making four separate points. The marking scheme identifies four key words or phrases that must be present for you to score the 4 marks. The total marks per question vary, up to 30 marks.

In Unit Test 1, it is rare to find one question concentrating on one topic. The questions here could represent two or three structured questions from separate papers. The important point is that if you score well in each question, then you can be confident that you understand the topic.

Record your scores and convert them to a percentage and a grade. As a rough guide, you need to score 80% to achieve grade A, 70% for grade B and 60% for grade C.

## Examiner's comments

Grade-A answers are provided for all questions. These are followed by examiner's comments (preceded by the icon *e*), which point out common errors and where a Grade C candidate is likely to lose marks. They also suggest alternative answers that are acceptable and ways of remembering key points.

# Atomic structure and mass spectrometry

**(a) Describe, in terms of charge and mass, the properties of protons, neutrons and electrons. Explain fully how these particles are arranged in an atom of $^{20}_{10}$Ne.** (6 marks)

**(b) Two other isotopes of neon exist, $^{21}_{10}$Ne and $^{22}_{10}$Ne. Explain the meaning of the term 'isotope'.** (2 marks)

**(c) The simplified mass spectrum of naturally occurring neon is shown below.**

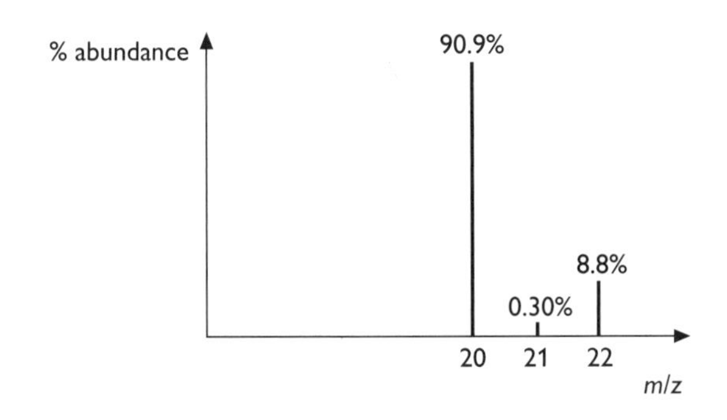


**Assuming that all the peaks are produced by ions with a single positive charge, what is the relative atomic mass of neon?** (3 marks)

**(d) Explain how the three isotopes of neon can be separated in a mass spectrometer by describing the various parts of the mass spectrometer and by discussing the principles of operation of each part.** (14 marks)

**(e) Another element was analysed in the mass spectrometer and the following data were produced.**

| ***m/z* value** | 134 | 135 | 136 | 137 | 138 |
|---|---|---|---|---|---|
| **Relative intensity** | 1 | 2.5 | 3.5 | 5.5 | 33 |

**Use the data to calculate the average relative atomic mass and identify the unknown element.** (3 marks)

**Total: 28 marks**

■ ■ ■

## Grade-A answer to Question 1

**(a)** The properties of charge and mass are summarised in the following table.

question



| Particle | Relative mass | Relative charge |
|---|---|---|
| Proton | 1 | +1 |
| Neutron | 1 | 0 |
| Electron | 1/1840 | −1 |

✓ ✓ ✓

The nucleus contains ten protons and ten neutrons ✓ and this is surrounded by electrons arranged in (energy) levels ✓ with an electron arrangement of $1s^2\,2s^2\,2p^6$ ✓.

Do not be afraid to summarise information in tables. The mass of an electron can be described as negligible.

**(b)** Isotopes are atoms with the same atomic number ✓ but a different mass number due to differing numbers of neutrons ✓.

You need to learn definitions so that they are presented concisely. 'Same proton number' instead of 'same atomic number' is also acceptable. A grade-C candidate might only get 1 mark here by not including the word 'atom' in the first phrase.

**(c)** $A_r = \frac{(20 \times 90.9) + (21 \times 0.3) + (22 \times 8.8)}{100}$ ✓ ✓ $= 20.179 = 20.2$ ✓

When calculating the average relative atomic mass, make sure that you show your working and that you express your answer to three significant figures. A Grade-C candidate is likely to lose a mark here, by giving the answer to three decimal places rather than to three significant figures.

**(d)** The gaseous ✓ sample of neon is introduced into the apparatus. An electron gun ✓ is used to bombard the atoms with electrons and this knocks out another electron ✓ from each atom to produce positive ions ✓ (cations).

The cations are accelerated ✓ by an electric field ✓ and passed through a slit so that they are focused into a beam of ions ✓.

The ions are deflected ✓ by passing them through a magnetic field ✓. The amount of deflection depends on the mass-to-charge ($m/z$) ratio ✓ of each ion. The ion with the smallest $m/z$ value is deflected the most ✓.

The ions then pass through the apparatus to the detector ✓. When the ions hit the detector, a small current ✓ is produced, which is amplified and sent to a computer where it is recorded and displayed ✓ as a chart. The magnetic or electric field can be adjusted to collect ions of different $m/z$ ratios ✓.

There are 15 scoring points ticked in this answer, but only 14 marks are available. This is quite common on longer questions and you need to score 14 of the 15 points to achieve full marks. Divide your answer into four short paragraphs, include around four key points in each one and do not include too much detail. Your answer should be in short sentences because the mark scheme will be summarised as key statements or key words. One equation that could be included to describe the ionisation process is:

$$Ne(g) + e^- \rightarrow Ne^+(g) + 2e^-$$

This would score 3 marks, because it implies gaseous atoms, electron loss and production of positive ions.

**(e)** $A_r = \dfrac{(134 \times 1) + (135 \times 2.5) + (136 \times 3.5) + (137 \times 5.5) + (138 \times 33)}{45.5}$ ✓

$= 137.4725275$

$= 137$ ✓

The unknown element is barium ✓.

A common mistake is to divide by 100 rather than by the total intensity (45.5). Doing this gives the incorrect answer 62.6 and identifies the element as copper. A Grade-C candidate may divide by 100 and/or give the answer to more than three significant figures.

# Electron arrangements and ionisation energies

**(a) The first ionisation energies of the elements sodium to silicon are given in the table below.**

| **Element** | Na | Mg | Al | Si |
|---|---|---|---|---|
| | | | | |

**(i) Explain the meaning of the term 'first ionisation energy'.** (3 marks)

**(ii) Explain why magnesium has a higher first ionisation energy than sodium.** (3 marks)

**(iii) State the full electron configurations of magnesium and aluminium.** (2 marks)

**(iv) Explain why the first ionisation energy of aluminium is less than that for magnesium.** (2 marks)

**(v) Predict a value for the first ionisation energy of phosphorus and explain how you estimated your answer.** (2 marks)

**(vi) State the full electron configurations of phosphorus and sulfur.** (2 marks)

**(vii) Explain why the first ionisation energy of sulfur is less than that of phosphorus.** (2 marks)

**(b) Write two separate equations, including state symbols, to describe the first and second ionisation energies of barium. Explain why the second ionisation energy of barium is higher than the first ionisation energy and why the third ionisation energy of barium is significantly higher than the second.** (7 marks)

**(c) Explain why the second ionisation energy of sodium is higher than the first ionisation energy of neon.** (3 marks)

**Total: 26 marks**

■ ■ ■

## Grade-A answer to Question 2

**(a) (i)** The energy change when 1 mole ✓ of gaseous atoms ✓ is converted into 1 mole of gaseous cations, with each atom losing an electron ✓.

*e* Try to be concise with your definition — remember three key points for 3 marks. It is worth including the general equation $X(g) \rightarrow X^{+}(g) + e^{-}$, because, if your definition is not clear, this would be given some credit.

**(ii)** Magnesium has more protons in the nucleus ✓, but the amount of shielding is the same ✓, and so the attraction for the outer electrons is greater ✓.

An alternative answer is that magnesium has a higher nuclear charge and the electrons are entering the same shell, so there is greater attraction for the outer electrons. The examiner is looking for an understanding of the concepts and although key words are written in mark schemes, other words are acceptable. A grade-C candidate may lose a mark here by not mentioning *outer* electrons.

**(iii)** Magnesium is $1s^2 2s^2 2p^6 3s^2$ ✓ and aluminium is $1s^2 2s^2 2p^6 3s^2 3p^1$ ✓.

Sometimes students give the answer [Ne]$3s^2$ for magnesium, but as this is not the *full* electron configuration, it is not acceptable. You must also remember to use *superscript* to show the number of electrons — $1s_2\, 2s_2\, 2p_6\, 3s_2$ is definitely not acceptable.

**(iv)** The electron in aluminium is removed from a $3p$ sub-level ✓, which is higher in energy ✓ than the $3s$ sub-level.

Many answers discuss the shielding of the $p$ sub-level by the $s$ sub-level, which is not what the examiner is looking for.

**(v)** The value for phosphorus will be about 995 kJ $mol^{-1}$ ✓. There is a regular increase from Al to P, so assuming the difference between Al and Si is the same as between Si and P, 209 + 786 = 995 ✓.

A grade-C candidate may not get the second mark. Predictive questions are quite common and you could be presented with a graph to complete, rather than numbers as here. These are easy questions, provided you remember that the pattern for both period 2 and period 3 is 2, 3, 3 followed by a big drop.

**(vi)** Phosphorus is $1s^2 2s^2 2p^6 3s^2 3p^3$ ✓ and sulfur is $1s^2 2s^2 2p^6 3s^2 3p^4$ ✓.

Remember to give the *full* electron configurations of the phosphorus and sulfur atoms. A useful way of showing whether one of the $p$ orbitals contains a pair of electrons is to write out the electron configurations to show the $3p$ orbitals as $p_X$, $p_Y$ and $p_Z$. So phosphorus is $1s^2 2s^2 2p^6 3s^2 3p^1{}_X 3p^1{}_Y 3p^1{}_Z$ and sulfur is $1s^2 2s^2 2p^6 3s^2 3p^2{}_X 3p^1{}_Y 3p^1{}_Z$. This would help the explanation in the next part of the question. A grade C candidate is unlikely to do this.

**(vii)** Sulfur contains a pair of electrons in one of the orbitals of the $3p$ level ✓ — [↑↓] [↑] [↑] — and this pair suffers repulsion, making the electron easier to remove compared with phosphorus, which has three unpaired electrons in the three orbitals — [↑] [↑] [↑] ✓.

A common phrase used by candidates is 'spin-pair repulsion'. However, to convince the examiner that you understand the concept, you need to include more detail. A grade-C candidate is unlikely to give a clear explanation and will probably gain only 1 mark.

**(b)** The first ionisation energy:

$Ba(g) \rightarrow Ba^+(g) + e^-$ ✓✓

The second ionisation energy:

$$Ba^{+}(g) \rightarrow Ba^{2+}(g) + e^{-}$$ ✓

The second ionisation energy is higher than the first because there is still the same number of protons ✓ and the ion is now positive and attracts the remaining electrons more strongly ✓.

The third ionisation energy is significantly higher than the second, because the electron is now removed from a new inner energy level, closer to the nucleus ✓, which experiences less shielding and so there is more attraction ✓.

*e* To gain the third mark in this question, you need to include all the correct state symbols. A common mistake in explaining the increase in successive ionisation energies is to state that the attraction is greater because the nuclear charge or the number of protons is increasing, when in fact they are constant. Another easy way to lose a mark is to forget to mention 'more attraction'.

**(c)** The electron configuration of neon is $1s^2\,2s^2\,2p^6$ and that of a sodium ion is $1s^2\,2s^2\,2p^6$. Both ionisations involve removing an electron from the $2p$ level ✓, but sodium has 11 protons in its nucleus, whereas neon has only ten ✓. The higher nuclear charge means a greater attraction to the electron in the $2p$ level ✓.

*e* The final question is difficult and would be used to discriminate at the top end of the ability range. When discussing questions of this type, always look to see how many protons are present in the nucleus, how many electrons surround the nucleus and the energy level from which the electron is removed.

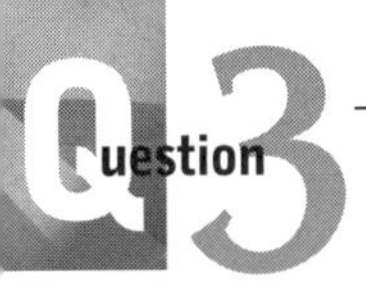




# Amount of substance (I)

**The gas constant $R = 8.31\ J\ K^{-1}\ mol^{-1}$.**

**(a) Define the term 'relative atomic mass'.** (2 marks)

**(b) (i) Define the terms 'empirical formula' and 'molecular formula'.** (2 marks)

**(ii) An organic compound containing 54.5% carbon, 36.4% oxygen and 9.1% hydrogen is analysed in a mass spectrometer and found to have a relative molecular mass of 88. Calculate the empirical formula and deduce the molecular formula of the compound. Write an equation for the complete combustion of this compound to produce carbon dioxide and water.** (6 marks)

**(c) (i) State the ideal gas equation and show how it can be used to calculate the relative molecular mass of a gas.** (3 marks)

**(ii) Use this expression to calculate the relative molecular mass of propane. 1.665 g of propane occupies a volume of 1.00 $dm^3$ at 323 K and 101 kPa.** (3 marks)

**(d) Calculate the volume of carbon dioxide that would be evolved when 4.00 g of sodium carbonate reacts completely with excess dilute hydrochloric acid at 30°C and 100 kPa.**

$$Na_2CO_3 + 2HCl \rightarrow 2NaCl + CO_2 + H_2O$$ (5 marks)

**(e) In the industrial extraction of iron, iron(III) oxide is reduced using carbon monoxide. How many tonnes of iron can be obtained from the complete reduction of 79.8 tonnes of iron(III) oxide with excess carbon monoxide?**

$$Fe_2O_3 + 3CO \rightarrow 2Fe + 3CO_2$$ (3 marks)

**Total: 24 marks**

■ ■ ■

## Grade-A answer to Question 3

**(a)** The relative atomic mass is the average mass of an atom of an element ✓ relative to $\frac{1}{12}$ the mass of an atom of carbon-12 ✓, which has a mass of 12.00 atomic mass units.

*e* You need to be able to recall this definition precisely. You could express it as a formula. A common error is to omit reference to an *atom* of an element and so lose the first mark.

**(b) (i)** The empirical formula is the simplest whole number ratio of atoms of each element in a compound ✓. The molecular formula is the actual number of atoms of each element in a compound ✓.

*e* Make sure that you learn all your definitions. In many questions in this unit, marks can be scored for recalling simple definitions.

**(ii)**

| | Carbon | Hydrogen | Oxygen | |
|---|---|---|---|---|
| % composition | 54.5 | 9.1 | 36.4 | |
| $A_r$ | 12.0 | 1.0 | 16.0 | ✓ |
| Divide by $A_r$ | 4.54 | 9.1 | 2.275 | ✓ |
| Divide by smallest | 2 | 4 | 1 | |

The 2:4:1 ratio shows that the empirical formula is $C_2H_4O$ ✓. The compound has a relative molecular mass of 88, and a relative empirical mass of 44. Therefore, its molecular formula is $C_4H_8O_2$ ✓.

$$C_4H_8O_2 + 5O_2 \rightarrow 4CO_2 + 4H_2O \text{ ✓ ✓}$$

*e* A common error in the first step of the calculation is to divide by the atomic number rather than $A_r$. This is a chemical error and so there will not be consequential marking.

The specification requires you to be able to write equations for unfamiliar reactions. A common mistake is to try to balance the equation without including oxygen on the left-hand side.

**(c) (i)** $pV = nRT$ ✓ and $n = \frac{\text{mass}}{M_r}$ ✓

So, $pV = \frac{mRT}{M_r}$

Rearranging gives:

$$M_r = \frac{mRT}{pV} \text{ ✓}$$

**(ii)** $M_r = \frac{1.665 \times 8.31 \times 323}{101 \times 10^3 \text{ ✓} \times 1.00 \times 10^{-3} \text{ ✓}} = 44.25 = 44.3$ ✓

*e* In part (i), it is essential to show the three stages involved in deriving the final expression. In part (ii) you are advised to show your working, just in case you make a careless slip. You should always convert kPa to Pa and $dm^3$ to $m^3$. A grade-C candidate is likely to lose the mark for the latter conversion.

**(d)** moles of $Na_2CO_3 = \frac{\text{mass}}{M_r}$

$= \frac{4.00}{106.0}$ ✓ $= 0.0377$ ✓

moles of $CO_2 = 0.0377$ ✓

$$V = \frac{nRT}{p}$$

$$= \frac{0.0377 \times 8.31 \times 303}{100 \times 10^3} \checkmark = 9.49 \times 10^{-4}\,\text{m}^3 \checkmark \text{ (or } 0.949\,\text{dm}^3 \text{ or } 949\,\text{cm}^3\text{)}$$

There are four stages in this calculation. The first mark is for the $M_r$ and the second is for the number of moles of $Na_2CO_3$. To achieve the third mark, you need to realise that the ratio of $Na_2CO_3 : CO_2$ is 1:1, as shown in the equation. The fourth mark is for using the rearranged ideal gas equation. Common mistakes include not converting °C to K (by adding 273) and not converting kPa to Pa. The final answer will only gain a mark if it is given to the correct number of significant figures and includes the units. A grade-C candidate is likely to score only the first 3 marks.

**(e)** $Fe_2O_3 \rightarrow 2Fe$

159.6 g ✓ → 2(55.8) = 111.6 g ✓

79.8 g → 55.8 g

79.8 tonnes → 55.8 tonnes ✓

This reacting masses question is relatively easy compared with some of the other calculations in this section. Marks are awarded for calculating the $M_r$ of the compounds, realising that it is a 1:2 ratio in the equation and scaling down to produce the correct answer, which must include the correct units.

# Amount of substance (II)

**(a) (i) Define the term 'relative molecular mass'.** (2 marks)

**(ii) The number of molecules in 1 mole of carbon dioxide is $6.023 \times 10^{23}$. What is the name given to this number of molecules?** (1 mark)

**(b) Magnesium reacts with dilute hydrochloric acid to produce magnesium chloride and hydrogen:**

$$Mg(s) + 2HCl(aq) \rightarrow MgCl_2(aq) + H_2(g)$$

**Calculate the volume of 0.500 mol dm$^{-3}$ hydrochloric acid needed to react completely with 1.50 g of magnesium ribbon.** (3 marks)

**(c) Magnesium carbonate decomposes to produce magnesium oxide and carbon dioxide:**

$$MgCO_3(s) \rightarrow MgO(s) + CO_2(g)$$

**(i) Calculate the maximum mass of magnesium oxide that could be formed when 10.0 g of magnesium carbonate is decomposed completely.** (3 marks)

**(ii) In an experiment involving the decomposition of magnesium carbonate, only 4.05 g of magnesium oxide was obtained. Calculate the percentage yield.** (2 marks)

**(d) You are provided with two separate solutions of 0.100 mol dm$^{-3}$ sulfuric acid.**

**(i) Calculate the mass of zinc needed to completely react with 50.0 cm$^3$ of a 0.100 M solution of sulfuric acid.**

$$Zn(s) + H_2SO_4(aq) \rightarrow ZnSO_4(aq) + H_2O(l)$$ (3 marks)

**(ii) Calculate the volume of 0.200 mol dm$^{-3}$ sodium hydroxide solution needed to react completely with 50.0 cm$^3$ of 0.100 mol dm$^{-3}$ solution of sulfuric acid.**

$$2NaOH(aq) + H_2SO_4(aq) \rightarrow Na_2SO_4(aq) + H_2O(l)$$ (2 marks)

**(e) Carbamide, $NH_2CONH_2$, is used as a fertiliser, in animal feeds, in paper processing and in a range of manufacturing processes. Its manufacture can be represented by the overall equation:**

$$CO_2 + 2NH_3 \rightarrow NH_2CONH_2 + H_2O$$

**Calculate the percentage atom economy for the production of carbamide.** (2 marks)

**Total: 18 marks**

■ ■ ■

## Grade-A answer to Question 4

**(a) (i)** Relative molecular mass, $M_r$, is the mass of a molecule ✓ compared to $\frac{1}{12}$ the mass of an atom of carbon-12 ✓, which has a mass of 12.00 atomic mass units.

**(ii)** $6.023 \times 10^{23}$ is the Avogadro constant ✓.

You can give the definition as a sentence or express it as a formula. 1 mole of any substance contains the same number ($6.023 \times 10^{23}$) of particles. This number is the Avogadro constant.

**(b)** moles of magnesium $= \dfrac{1.50}{24.3} = 0.06173$ ✓

moles of HCl $= 2 \times 0.06173 = 0.1235$ ✓

$$n = \frac{\text{concentration} \times \text{volume}}{1000}$$

Therefore, $0.1235 = \dfrac{0.500 \times \text{volume}}{1000}$

Rearranging gives:

$$\text{volume} = \frac{0.1235 \times 1000}{0.500} = 247\ \text{cm}^3\ ✓$$

The correct final answer, with the appropriate number of significant figures and units, would score 3 marks. However, it is essential to show your working, because careless mistakes can be made. If you make a slip in calculating the number of moles of magnesium, you can still score the next 2 marks in each calculation, provided the working is correct.

**(c) (i)** $MgCO_3(s) \rightarrow MgO(s) + CO_2(g)$

moles of magnesium carbonate $= \dfrac{10.0}{84.3} = 0.1186$ ✓

moles of magnesium oxide = 0.119 ✓

mass of magnesium oxide = moles $\times M_r = 0.1186 \times 40.0 = 4.75$ g ✓

**(ii)** % yield $= \dfrac{\text{mass obtained} \times 100}{\text{calculated mass}}$

$= \dfrac{4.05 \times 100}{4.75}$ ✓ $= 85.263 = 85.3\%$ ✓

The % yield calculation is dependent on a mass of magnesium oxide being calculated in (i). Without this, a % yield cannot be found, so always attempt the first part in any calculation of this type.

**(d) (i)** moles of sulfuric acid $= \dfrac{\text{concentration} \times \text{volume}}{1000}$

$= \dfrac{0.100 \times 50.0}{1000} = 5 \times 10^{-3}$ ✓

moles of zinc $= 5 \times 10^{-3}$ ✓

mass of zinc $= 5 \times 10^{-3} \times 65.4 = 0.327$ g ✓

**(ii)** moles of sulfuric acid $= 5 \times 10^{-3}$

1:2 ratio in the equation

$\therefore$ moles of NaOH $= 1 \times 10^{-2}$ ✓

$$\text{volume of NaOH required, } V, = \frac{\text{moles} \times 1000}{\text{concentration}}$$

$$= \frac{1 \times 10^{-2} \times 1000}{0.200} = 50.0\,\text{cm}^3 \checkmark$$

*e* Remember the three key stages in this type of calculation:

- Calculate moles of acid using $\frac{\text{concentration} \times V}{1000}$
- Use the ratio in the equation, i.e. 1:1 or 1:2.
- Use $n = \frac{\text{mass}}{M_r}$ if a solid is involved or $n = \frac{\text{concentration} \times V}{1000}$ if a solution is involved.

**(e)** $$\%\text{ atom economy} = \frac{\text{mass carbamide}}{\text{total mass reactants}} \times 100$$

$$= \frac{60.0}{44.0 + 34.0} \times 100 \checkmark = 76.9\% \checkmark$$

*e* Actual masses do not have to be given in questions about percentage atom economy; the required masses are obtained from the formulae given. Be sure to use the correct number of moles of reactants — in this case, 1 mol $CO_2$ and 2 mol $NH_3$.

# Types of bonding

*Note*: You should revise bonding types and shapes of molecules before attempting this question.

**(a) Using *only* the elements *potassium* and *fluorine* or *any combination* of these two elements, give an example and then explain the formation and principle features of:**

**(i) ionic bonds** (4 marks)

**(ii) covalent bonds** (4 marks)

**(iii) metallic bonds** (4 marks)

**(b) Name and sketch the shape of water ($H_2O$) and ammonia ($NH_3$) molecules, indicating the bond angles and the presence of any lone pairs.** (8 marks)

**(c) Molecules of water and ammonia can both accept an $H^+$ ion by forming a coordinate bond. Define the term coordinate bond, name and sketch the shape, including the bond angles, of the new species formed in each case and comment on the relative strengths of all the N–H bonds in the final species.** (10 marks)

**Total: 30 marks**

■ ■ ■

## Grade-A answer to Question 5

**(a) (i)** An ionic bond is formed between potassium and fluorine to give KF ✓. Potassium ($1s^2\,2s^2\,2p^6\,3s^2\,3p^6\,4s^1$) loses an electron to become $K^+$ ✓. The electron is transferred ✓ to the fluorine atom. Fluorine ($1s^2\,2s^2\,2p^5$) gains an electron to become a fluoride ion, $F^-$ ✓. Electrostatic forces of attraction hold the oppositely charged ions together ✓ (in a giant, three-dimensional, ionic lattice ✓.

**(ii)** A covalent bond is formed between two fluorine atoms to give $F_2$ ✓. Both fluorine atoms ($1s^2\,2s^2\,2p^5$) need to gain an extra electron ✓ and this is achieved by sharing an electron pair ✓. Each of the atoms contributes one electron to the shared pair ✓.

**(iii)** A metallic bond is present in the element potassium, K ✓. Potassium atoms lose control of their outer electron to form $K^+$ cations ✓. The electrons that are lost form a sea of delocalised electrons ✓. The metal is a three-dimensional lattice of cations in a closely packed ✓ arrangement held together by their mutual attraction for the delocalised electrons ✓.

*e* This answer offers 14 scoring points. However, only 12 marks can be scored in total. You should structure your answer using the question and the mark allocation given. There are three types of bonding to discuss and the question asks for an example in each case. The mark scheme will usually give 1 mark for each example and 3 marks for each explanation.

**(b)**

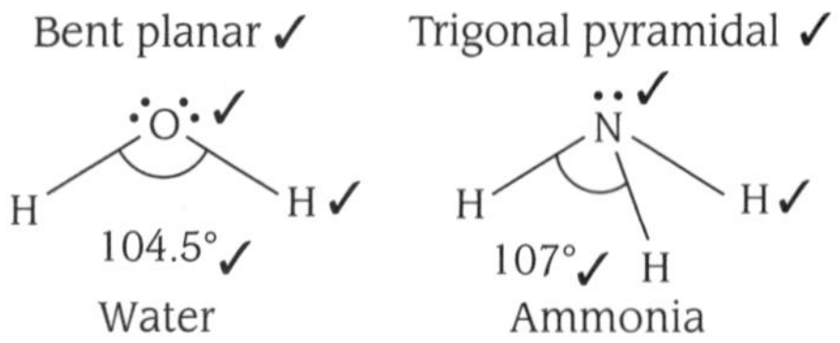


The question and the mark allocation indicate how to structure your answer. When reading through questions in the exam, you should highlight the key words in the question — in this case, 'name', 'sketch', 'bond angles' and 'lone pairs'. These four key points equate to 4 marks and as you have to discuss two molecules, it is clear how the 8 marks are awarded.

**(c)** A coordinate bond is defined as a shared pair of electrons ✓ in which both electrons have originated from one atom ✓.

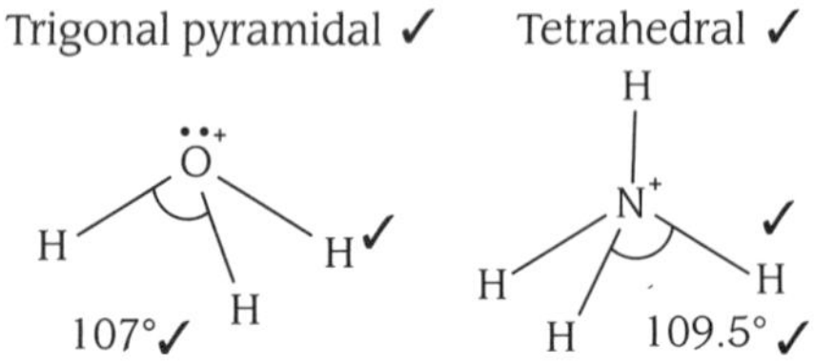


The four N–H bonds in the final species ($NH_4^+$, ammonium ion) are all exactly the same strength ✓ because they involve the sharing of an electron pair ✓, i.e. a coordinate bond is the same as any other covalent bond.

Repeat the same procedure here. Two ions each require a name, a sketch and the bond angles, producing 6 marks. 2 marks for the definition and 2 marks for the final explanation give a total of 10 marks.

**It is common to see a question that requires the understanding of more than one topic. In this question on bonding, a detailed knowledge of the basic principles of shapes of molecules is also required. This question demonstrates how to use the requirements and the mark allocation to structure your answer.**

# Bond polarity

**(a) (i) Define the term 'electronegativity'.** (2 marks)

**(ii) Using suitable simple molecules as examples, explain the terms 'non-polar covalent bond' and 'polar covalent bond'.** (7 marks)

**(b) The structures of chloromethane and tetrachloromethane are shown below:**

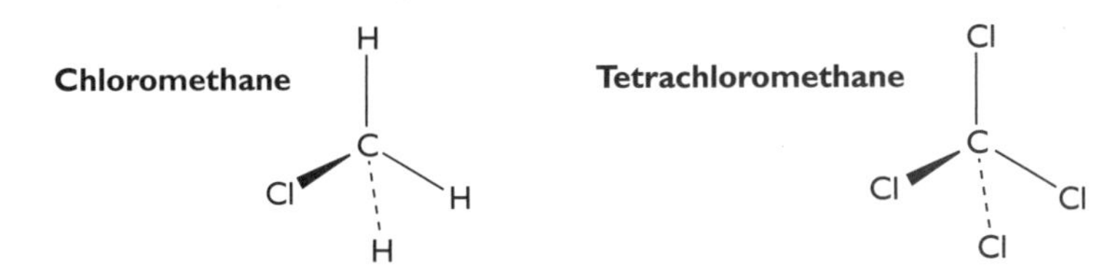


**Both molecules have four bonds arranged in a tetrahedral fashion around the carbon atom. In chloromethane, show the bond polarity of the carbon–chlorine bond by using the symbols δ+ and δ− to indicate the electron-rich and electron-deficient atoms. Suggest why the molecules of tetrachloromethane show no overall polarity.** (3 marks)

**Total: 12 marks**

■ ■ ■

## Grade-A answer to Question 6

**(a) (i)** Electronegativity is the ability of an atom to attract the electron density ✓ in a covalent bond ✓.

'The ability of an atom to attract an electron pair in a covalent bond' is also an acceptable answer. 'The ability to attract an electron' would fail to score — it is essential that you refer to the covalent bond. It is likely that a grade-C candidate would not score the second mark.

**(ii)** A covalent bond is a shared electron pair ✓. This electron pair is shared equally when both atoms have the same electronegativity ✓, so the bond is non-polar. An example is chlorine.

$$\mathrm{Cl}\overset{\times}{\underset{\times}{—}}\mathrm{Cl}$$ ✓

A polar covalent bond occurs when the electron pair is not shared equally ✓. In hydrogen chloride, for example, the chlorine is more electronegative than the hydrogen ✓, so the bond is polar. The chlorine is electron-rich and the hydrogen is electron-deficient ✓.

$$\overset{\delta+}{\mathrm{H}}—\overset{\times}{\underset{\bullet}{}}\overset{\delta-}{\mathrm{Cl}}$$ ✓

Use simple examples to illustrate your answer. Any diatomic molecule can be used to illustrate a non-polar bond, for example $H_2$, $O_2$, $F_2$, $Br_2$, $I_2$. Simple molecules that contain elements of different electronegativities can be used to illustrate a polar covalent bond — for example, $H_2O$ and $NH_3$. The mark for the electron-deficient and electron-rich sites can be scored by labelling the diagram with δ+ and δ−.

**(b)** The polarity of the carbon–chlorine bond in chloromethane is shown below.

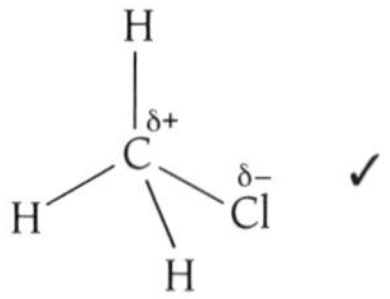

H
C δ+
Cl δ− ✓
H H

$CCl_4$ is a symmetrical ✓ molecule, so all the bond polarities cancel out ✓ to give zero polarity.

A grade-C candidate is likely to omit the last statement and so gain only 2 marks. Molecules that have polar bonds arranged in a symmetrical fashion have no overall polarity. Other examples are:

- $CO_2$, a linear molecule with no overall polarity
- $SO_2$, a bent planar molecule with an overall dipole

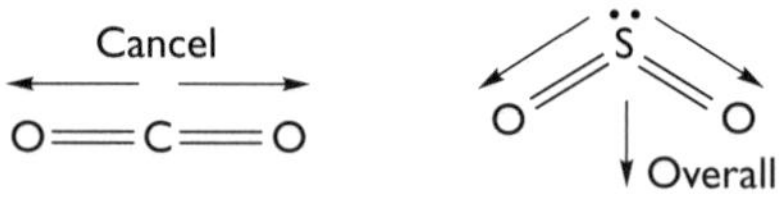

Cancel
O═C═O

S
O O
Overall

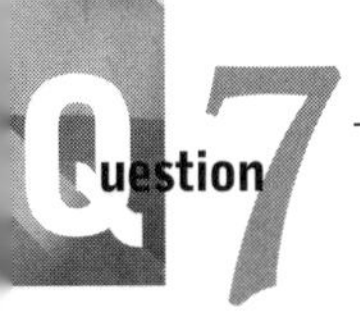

# Intermolecular forces and states of matter

**(a) There are three types of intermolecular force: induced dipole–dipole (van der Waals) forces, permanent dipole–dipole forces and hydrogen bonding. Using an example in each case, explain how each type of intermolecular force arises.** (12 marks)

**(b) (i) Explain why iodine exists as a solid and chlorine exists as a gas at room temperature and pressure.** (2 marks)

**(ii) Explain why solid iodine can be vaporised by gentle heating.** (2 marks)

**(c) Sketch a molecule of ammonia and include any lone pairs of electrons. Show how a hydrogen bond forms between this ammonia molecule and a neighbouring ammonia molecule. Indicate the polarity of any atoms involved in the hydrogen bond with the symbols $\delta+$ and $\delta-$.** (4 marks)

**(d) Which of the molecules shown below does not exhibit hydrogen bonding? Give a brief reason for your answer.** (2 marks)

$H_3C$–O–H — Methanol

$H_3C$–C(=O)–H — Methanal

$H_3C$–C(=O)–OH — Methanoic acid

**(e) Indicate which compound in each of the following pairs of substances has the higher boiling point. Give a brief reason for your answer.**

**(i) $CH_4$ and $SiH_4$** (2 marks)

**(ii) $C_3H_8$ and $CH_3Cl$** (3 marks)

**(iii) $H_2O$ and $H_2S$** (3 marks)

**Total: 30 marks**

■ ■ ■

## Grade-A answer to Question 7

**(a)** Induced dipole–dipole (van der Waals) forces attract iodine molecules ✓ to each other. Iodine is a non-polar molecule ✓, but the electron distribution can become asymmetrical (unequal) leading to a temporary dipole ✓. This induces an opposite dipole in the adjacent molecule, which is therefore attracted to the first molecule ✓.

Permanent dipole–dipole forces attract hydrogen chloride molecules ✓ to each other. It is a polar molecule because permanent charge separation exists between the hydrogen and chlorine atoms ✓. Chlorine is more electronegative than hydrogen, so chlorine is electron-rich and hydrogen is electron-deficient ✓. One dipole attracts the opposite dipole on a neighbouring molecule ✓.

Hydrogen bonding is found between molecules such as water ✓. The hydrogen atom is attached directly to a small electronegative oxygen atom ✓, and the hydrogen becomes so electron-deficient ✓ that it attracts a lone pair of electrons from the oxygen of a neighbouring molecule ✓.

*e* For each type of intermolecular force there is 1 mark for the example and 3 marks for the explanation. You could include *labelled* diagrams to improve your chances of scoring all the points. The examiner is looking for an understanding of the concept and will look for certain key points in a description or implied in a diagram. A grade-C candidate is likely to gain the marks for the examples but lose some of the explanation marks because of insufficient detail.

**(b) (i)** A molecule of iodine is larger and contains many more electrons than a molecule of chlorine ✓, so it has much stronger induced dipole–dipole (van der Waals) forces ✓.

**(ii)** Iodine can be vaporised by gentle heating because the induced dipole–dipole (van der Waals) forces between the molecules are weak ✓ and require little energy to overcome them ✓.

*e* Questions on states of matter are common and most require only simple answers. Most candidates would find this question relatively straightforward.

**(c)**

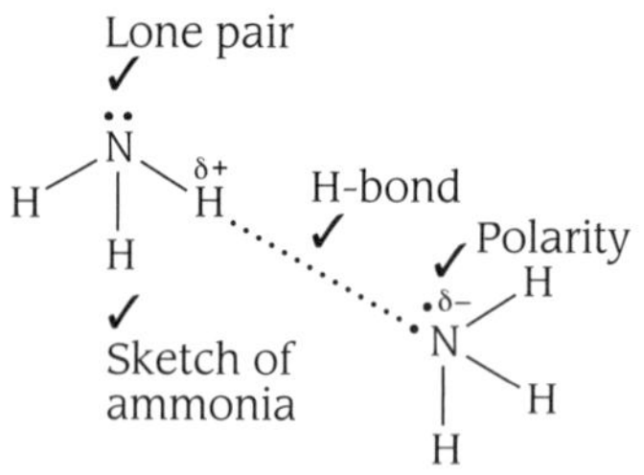

Lone pair ✓

H-bond ✓

Polarity ✓

$\delta+$ H ····· $\delta-$ N

Sketch of ammonia ✓

*e* The examiner requires a clear sketch of ammonia that shows three N–H bonds and one lone pair of electrons on the nitrogen. The shape of the molecule is pyramidal with bond angles of 107°. However, the geometry of ammonia is not being tested, so there are no marks available for this on your sketch. You must show the hydrogen bond (usually indicated by a dotted line) between the hydrogen of the first molecule and the lone pair on the nitrogen of the second molecule. The final mark is for showing the electron-deficient hydrogen and the electron-rich nitrogen.

**(d)** Methanal does not exhibit hydrogen bonding ✓ because the hydrogen atom must be attached directly to the electronegative oxygen atom ✓.

*e* This is an application question that tests understanding of the concept of hydrogen bonding. The marks are likely to be scored by good candidates only; others may be put off by compounds that they may not have met before. The key point here is that hydrogen must be attached directly to a small electronegative atom, such as oxygen, to cause hydrogen bonding between molecules.

**(e) (i)** $SiH_4$ ✓ has the higher boiling point because it is a larger molecule with more electrons and hence stronger induced dipole–dipole (van der Waals) forces ✓.

**(ii)** $CH_3Cl$ ✓ has a higher boiling point because it is a polar molecule, and the permanent dipole–dipole forces ✓ are stronger than the induced dipole–dipole (van der Waals) forces found between $C_3H_8$ molecules ✓.

**(iii)** $H_2O$ ✓ has a higher boiling point because the hydrogen bonding ✓ is stronger than the dipole–dipole forces found in hydrogen sulfide ✓.

*e* If you understand the basic principles of intermolecular forces, you should be able to apply them to any molecule. The order of strength of intermolecular forces is induced dipole–dipole (van der Waals) forces < permanent dipole–dipole forces < hydrogen bonding. To recognise the type of intermolecular forces present, remember the following three key points:

- All covalent molecules show induced dipole–dipole (van der Waals) forces.
- To show permanent dipole–dipole forces, there must be atoms of elements of differing electronegativity present.
- To show hydrogen bonding, hydrogen must be present and directly attached to nitrogen, oxygen or fluorine.

# Types of crystal

**A crystal is a solid with particles organised in a regular structure. Consider the following crystal structures:**

- **sodium chloride**
- **iodine**
- **diamond**
- **graphite**

**(a) Draw simple sketches to illustrate the bonding and arrangement of particles found in these crystal structures.** (8 marks)

**(b) State the type of particle found in each type of crystal and the force of attraction that holds these particles together.** (8 marks)

**(c) Name the crystal structure (sodium chloride, iodine, diamond or graphite) that exhibits each of the following properties and give a brief explanation of this property in terms of structure and bonding:**

**(i) sublimes on gentle warming**

**(ii) conducts electricity in the solid state**

**(iii) very hard, with an extremely high melting point**

**(iv) conducts electricity only when molten or in solution** (12 marks)

**(d) State one other type of crystal structure and give an example.** (2 marks)

**Total: 30 marks**

## Grade-A answer to Question 8

**(a)**

Sodium chloride

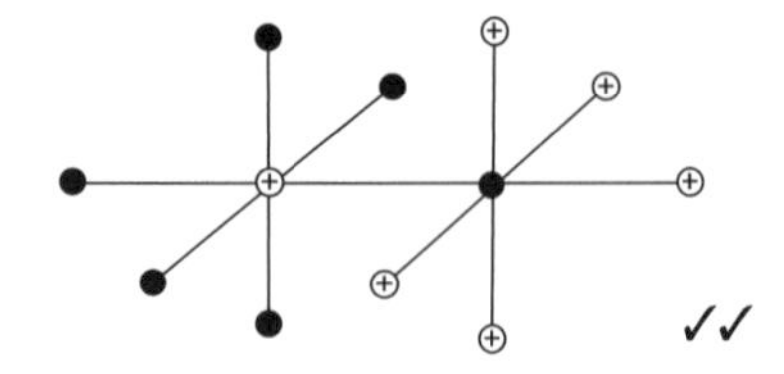


Iodine

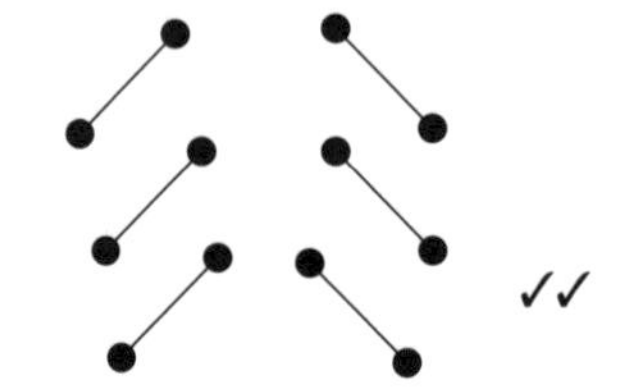

Diamond

✓✓

Graphite

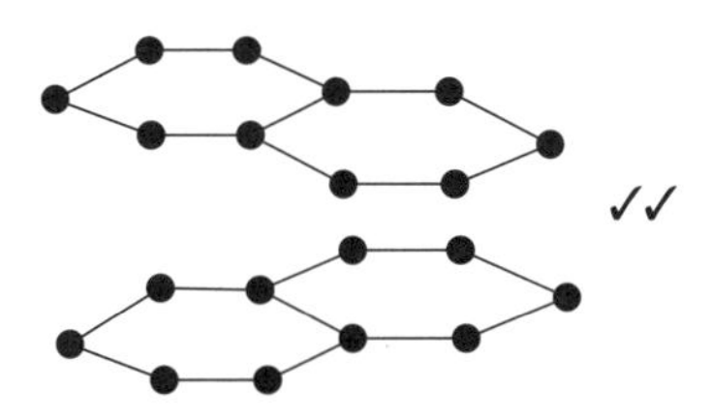

✓✓

The sketches do not have to be to the same standard as those in textbooks. They need to be good enough to convince the examiner that you understand the key features of each crystal lattice. The sodium chloride lattice needs to show *oppositely charged ions in a lattice* for the first mark and the idea of a *three-dimensional structure* for the second mark. An alternative acceptable sketch is shown below:

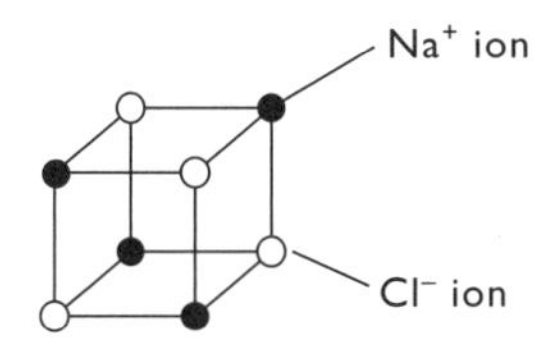


Iodine is a difficult structure to draw, but the simple sketch above is allowed and gains the first mark for the fact that iodine exists as diatomic molecules, and the second mark for showing a regular pattern. The structure of diamond must show a carbon attached to four other carbon atoms for the first mark, and, for the second mark, at least one more carbon attached to imply that diamond is macromolecular. You must be careful not to draw carbon atoms with more than four bonds. In graphite, the first mark is for implying that it is macromolecular by correctly linking together at least two hexagons. The second mark is for showing that the structure has different layers, so another hexagon needs to be drawn in a separate layer. A common mistake is to join the hexagons at the corners.

**(b)** Sodium chloride contains ions ✓. The oppositely charged ions are held together by strong electrostatic forces of attraction ✓.

Iodine is made up of simple molecules ✓, which are held together by induced dipole–dipole (van der Waals) forces ✓.

Diamond is a macromolecule made up of many carbon atoms ✓, linked together by strong covalent bonds ✓.

**question**



Graphite is another macromolecule made up of many carbon atoms arranged in layers ✓. There are strong covalent bonds between the carbon atoms in the layers and weak forces between the layers ✓.

*e* Diamond and graphite are macromolecules; the particles that make up the crystal are carbon atoms. For the final mark to be awarded, it is essential that you mention the weak forces between the layers in graphite.

**(c) (i)** Iodine sublimes on gentle heating ✓ because to separate the molecules the induced dipole–dipole (van der Waals) forces ✓ need to be overcome, and these are extremely weak ✓.

**(ii)** Graphite is the only crystal structure listed here that can conduct electricity in the solid state ✓, because some electrons (one from each carbon atom) are delocalised ✓ and are free to move between the layers ✓.

**(iii)** Diamond is hard with an extremely high melting point ✓ because all the carbon atoms are linked by covalent bonds ✓, which are strong ✓.

**(iv)** Sodium chloride only conducts when molten or in solution ✓. This is because the ions vibrate about a fixed position in the solid state ✓, but when molten or in solution, they become mobile ✓.

*e* Use the question and the mark allocation to structure your answer. There are 12 marks available and four properties to discuss. This means that, in each case, there is 1 mark for the example and 2 marks for the explanation. In the explanations, you must be careful how you word your answers. For example, for part (i) 'Iodine sublimes on gentle heating because to separate the *atoms* the induced dipole–dipole (van der Waals) forces need to be overcome, and these are weak' would not score because the induced dipole–dipole (van der Waals) forces do not exist between the iodine atoms. Another example of a potential slip is 'Sodium chloride only conducts when molten or in solution. This is because the *atoms* vibrate about a fixed position in the solid state, but when molten or in solution they become mobile'. This would also fail to score because sodium chloride is ionic and, therefore, atoms must *never* be mentioned.

**(d)** Another type of crystal is metallic ✓. An example is magnesium ✓.

*e* Any named metal is a suitable answer for the example of a metallic lattice. There are four types of lattice: ionic (sodium chloride), molecular (iodine), macromolecular (diamond or graphite) and metallic (magnesium). If you are asked to describe four types of crystal lattice, do not include both diamond and graphite in your answer.

# Shapes of molecules and ions



**(a) Nitrogen has an atomic number of 7 and a mass number of 14.**
**Consider the following nitrogen species:**

- **$NH_4^+$, ammonium ion**
- **$NH_3$, ammonia**
- **$NH_2^-$, amide ion**
- **$N^{3-}$, nitride ion**

**(i) State the electron configuration of the nitrogen atom and state the number of unpaired electrons in the atom.** (2 marks)

**(ii) State the number of protons, neutrons and electrons in a nitride ion.** (3 marks)

**(iii) Explain how an ammonia molecule could be converted into an ammonium ion and state the type of bond formed in this conversion.** (3 marks)

**(iv) Explain the differences in the H–N–H bond angles found in the ammonia molecule and the ammonium ion. Include in your answer a sketch, the name of the shape of each species, the bond angles and a reason for any difference.** (8 marks)

**(v) Use the electron pair repulsion theory to predict the shape of the amide ion. Include in your answer a sketch of the molecule, indicating any bond angles and lone pairs.** (3 marks)

**(b) Consider the following molecules or ions:**

- **$BF_3$**
- **$PCl_5$**
- **$PCl_3$**
- **$PF_6^-$**
- **$PCl_4^+$**

**For each of the molecules or ions, state its shape and give the values of the bond angle(s) found within the molecule or ion.** (11 marks)

**Total: 30 marks**

■ ■ ■

## Grade-A answer to Question 9

**(a) (i)** Nitrogen has the electron configuration $1s^2 2s^2 2p^3$ ✓. It has three ✓ unpaired electrons in each of the *p* orbitals, i.e. [↑][↑][↑].

*e* Common mistakes are writing 1s2 2s2 2*p*3 or $1s_2\ 2s_2\ 2p_3$. Neither of these is acceptable. Nitrogen has three electrons in the *p* sub-level arranged as unpaired electrons in separate orbitals, i.e. $2p_X^1\ 2p_Y^1\ 2p_Z^1$ and not $2p_X^2\ 2p_Y^1$.

**(ii)** The nitride ion, $N^{3-}$, has seven protons ✓, seven neutrons ✓ and ten electrons ✓.

The number of protons is equal to the atomic number and the number of neutrons is equal to the difference between the mass number and the atomic number. There is a charge of 3– on the nitride ion. Therefore, three additional electrons are present, making ten electrons.

**(iii)** The ammonia molecule contains a lone pair of electrons on the nitrogen atom ✓, so it can readily accept an $H^+$ ion ✓, by reacting with an acid such as HCl, forming a coordinate (dative covalent) bond ✓ between the nitrogen atom and the $H^+$ ion.

It is essential that you recognise the key features of coordinate bond formation, i.e. the presence of a lone pair of electrons on the nitrogen atom in ammonia and the ability to accept an $H^+$ ion.

**(iv)** Ammonia has four pairs of electrons, so is based on a tetrahedral structure. However, one of the electron pairs is a lone pair, so the H–N–H bond angle is reduced to 107° because lone pair/bond pair repulsion is greater than bond pair/bond pair repulsion ✓. When ammonia is converted to the ammonium ion, the lone pair becomes a bonding pair (the coordinate bond), so there are now four bonding pairs which repel each other equally ✓, giving bond angles of 109.5°.

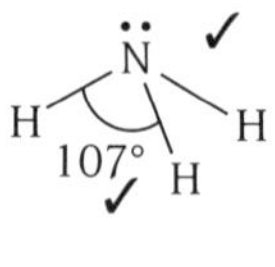

N ✓
H H
107° H
✓

Trigonal pyramidal ✓

H
N$^+$ ✓
H H
109.5° H
✓

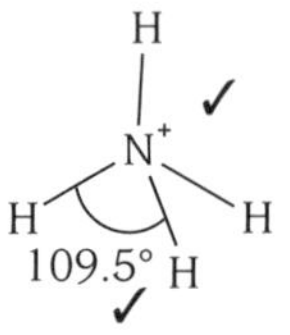

Tetrahedral ✓

In this question, the mark allocation can be used to determine the structure of your answer. You have to draw two sketches, each of which is labelled with the name of the shape and a bond angle, giving rise to 6 marks. The remaining 2 marks are for the explanation. The examiner is looking for an understanding of the key difference between the molecule and the ion, i.e. that both are based on four electron pairs but one has a lone pair present. It is this lone pair that causes the reduction in the bond angle.

**(v)** The nitrogen has five electrons in the outer shell and it gains two more electrons from the hydrogen in the formation of the amide ion. Add another electron because the ion is negative and this gives a total of eight electrons. Dividing by two gives four electron pairs, so the shape is based on a tetrahedral structure with a bond angle of 109.5°. However, two of the electron pairs are lone pairs, so the bond angle is reduced by about 5°, to give a bent planar molecule with a bond angle of 104.5°

H–N–H with two lone pairs on N ✓; bond angle 104.5° ✓ ✓

Amide ions are unfamiliar and some candidates will find this question intimidating. In this application question, you have to use the basic principles covered in the section on shapes of molecules. The question asks for a sketch, the bond angle and an indication of the number of lone pairs; these three key points are worth 3 marks. There are no marks for the name of the shape or for the reasoning behind the final answer. However, read through the answer again because this explanation is the thought process that you need to go through in order to arrive at the final answer. The answer is similar to deducing the shape of water molecules.

**(b)**

| Molecule or ion | Shape | Bond angle(s) |
|---|---|---|
| $BF_3$ | Trigonal planar ✓ | 120° ✓ |
| $PCl_5$ | Trigonal bipyramidal ✓ | 90° ✓ and 120° ✓ |
| $PCl_3$ | Trigonal pyramidal ✓ | 107° ✓ |
| $PF_6^-$ | Octahedral ✓ | 90° ✓ |
| $PCl_4^+$ | Tetrahedral ✓ | 109.5° ✓ |

This part question requires only the name of the molecular shape and the bond angles. No sketch is required and neither is an explanation of the basic principles. However, this question could easily be expanded to require a sketch or some explanation of how the shapes were deduced. If you score 11 marks here, you have a good understanding of the basic principles of the shapes of molecules and ions. There are 2 marks available for the two bond angles in $PCl_5$ — this is why five examples lead to 11 marks. If there is an unexpected additional mark, it is possible that a trigonal bipyramidal structure is one of the answers because this is the only one with two bond angles.

**Do not be surprised to see questions on other topics when answering a question on shapes of molecules. You will never see a 30-mark question purely on shapes of molecules and ions. It is essential to produce clear, labelled diagrams when drawing simple covalent molecules.**

# Periodicity

**(a) State the electron configurations of sodium and phosphorus and classify each element as an *s*-block or a *p*-block element. State the name and give the full electron configuration of a *d*-block element with an atomic number of less than 30.** (5 marks)

**(b) Use your knowledge of atomic structure and bonding to explain the following:**

**(i) the electrical conductivity of sodium is greater than that of phosphorus** (3 marks)

**(ii) the atomic radius of sodium is greater than that of phosphorus** (3 marks)

**(iii) the first ionisation energy of phosphorus is greater than that of sulfur** (3 marks)

**(iv) the melting point of sodium is lower than that of magnesium** (3 marks)

**(v) the melting point of phosphorus is much lower than that of silicon** (4 marks)

**(vi) the melting point of phosphorus is lower than that of sulfur** (3 marks)

**(c) Use your knowledge of the chemistry of period 3 to predict, with a brief explanation, the name of the element in period 2 that has:**

- **the largest atomic radius**
- **the largest first ionisation energy**
- **the highest melting point**

**A different element should be named for each physical property.** (6 marks)

**Total: 30 marks**

■ ■ ■

## Grade-A answer to Question 10

**(a)** Sodium is $1s^2\,2s^2\,2p^6\,3s^1$ ✓ and is an *s*-block ✓ element. Phosphorus is $1s^2\,2s^2\,2p^6\,3s^2\,3p^3$ ✓ and is a *p*-block ✓ element. Vanadium is a *d*-block element and the electron configuration is $1s^2\,2s^2\,2p^6\,3s^2\,3p^6\,3d^3\,4s^2$ ✓.

*e* This is a relatively straightforward question and most candidates will score the first 4 marks. Remember to give the full electron configurations. It is essential to use superscripts to show the number of electrons in each sub-level. In order to score the last mark in this question, you must relate the electron configuration to the stated *d*-block element. The atomic number must be less than 30, so any correct electron configurations for the elements scandium to copper would be accepted. A common answer would be iron: $1s^2\,2s^2\,2p^6\,3s^2\,3p^6\,3d^6\,4s^2$. Two common mistakes occur when writing out the electron configurations of copper and chromium. These two *d*-block elements are exceptions to the general pattern: copper is $1s^2\,2s^2\,2p^6\,3s^2\,3p^6\,3d^{10}\,4s^1$ and chromium is $1s^2\,2s^2\,2p^6\,3s^2\,3p^6\,3d^5\,4s^1$.

**(b) (i)** Sodium has metallic ✓ bonding. Therefore, because of the mobile electrons ✓, it is a good electrical conductor. Phosphorus ($P_4$) is a non-metal and so does not conduct electricity because the electrons are localised in the covalent bonds ✓ of the molecule.

**e** In order to gain full marks you must remember that phosphorus exists as a $P_4$ molecule. A grade-C candidate is unlikely to gain this final mark.

**(ii)** Sodium and phosphorus have the same amount of shielding ✓, but sodium has fewer protons in the nucleus ✓. This means that in sodium the nucleus has a smaller attraction for the outer electrons ✓ and so the atomic radius is larger.

**e** There are three key points to cover. The answer could read 'both have outer electrons in the same level, but phosphorus has a higher nuclear charge and so attracts the outer electrons more than sodium'. This covers the same three key points. Examiners are flexible when marking this answer, provided you can prove that you *understand* these three key points.

**(iii)** The first ionisations of phosphorus and sulfur both involve removing an electron from the 3*p* level. In phosphorus ($1s^2\,2s^2\,2p^6\,3s^2\,3p_X^1\,3p_Y^1\,3p_Z^1$) the electrons in the 3*p* level are all unpaired ✓, whereas in sulfur ($1s^2\,2s^2\,2p^6\,3s^2\,3p_X^2\,3p_Y^1\,3p_Z^1$) two of the electrons are paired ✓ and suffer repulsion, making it easier to remove an electron ✓.

**e** When answering questions on the drop in ionisation energy between group 5 and group 6 elements, always include the electron configuration of both elements. These electron configurations imply that you know that the electrons are in the 3*p* level; phosphorus has three unpaired electrons whereas sulfur has a pair of electrons in one of the 3*p* orbitals.

**(iv)** Sodium and magnesium are both metals and so have a lattice of cations held together by their attraction for the delocalised electrons ✓. The metallic bonding is stronger in magnesium because there is more attraction between the cations and delocalised electrons ✓. There is more attraction because the $Mg^{2+}$ cation has a higher charge than the $Na^+$ ion ✓; the $Mg^{2+}$ is smaller (so it has a higher charge density) and it has more electrons per cation holding the lattice together.

**e** The final sentence of the answer contains three reasons why the metallic bond is stronger and any one of these would score the final mark. The examiner is testing your understanding of the key points: the type of bonding present, the relative strength of the bonds and the reason for the difference in strength.

**(v)** Phosphorus is a simple covalent molecule ✓ with weak induced dipole–dipole (van der Waals) forces between the molecules ✓. Silicon is a macromolecule ✓ with strong covalent bonds between all the atoms ✓.

**e** When explaining melting points, it is essential that:

- you use the word 'weak' or 'strong' when discussing the forces between the particles
- you identify clearly the relevant particles and structures

**(vi)** Phosphorus and sulfur are both simple molecular solids ✓. However, $S_8$ is a larger molecule ✓ than $P_4$, so the induced dipole–dipole (van der Waals) forces between sulfur molecules are much stronger than between phosphorus molecules ✓.

*e* When discussing the melting points of simple covalent molecules, it is essential to give the formula of each molecule. The three molecules usually discussed are phosphorus, sulfur and chlorine; the order of melting point is related to the size of the molecules and, hence, the total number of electrons. Argon is made up of single atoms, so it has the lowest melting point in period 3. The order of melting points is $S_8 > P_4 > Cl_2 > Ar$.

**(c)** The element in period 2 with the largest atomic radius is lithium ✓. Lithium has the same amount of shielding as the other elements, but it has the fewest protons in its nucleus ✓.

The element in period 2 with the largest first ionisation energy is neon ✓. The outer electrons in neon have the same amount of shielding as in the other elements, but neon has the greatest number of protons in its nucleus and so has a greater attraction for the electron being removed ✓.

The element in period 2 with the highest melting point is carbon ✓. Carbon has a macromolecular structure, with strong covalent bonds between the atoms ✓.

*e* This question has 3 marks for the names of the elements given for each physical property. If the name is incorrect, then the explanation mark will not be awarded for that particular physical property. The important point here is that the candidate fully understands the concept of periodicity and that the patterns observed in period 2 are similar to those in period 3.

# Nomenclature and isomerism in organic chemistry

**(a) Two structural isomers of $C_6H_{14}$ are shown below:**

Hexane

$CH_3—CH_2—CH_2—CH_2—CH_2—CH_3$

2-methylpentane

$CH_3—CH(CH_3)—CH_2—CH_2—CH_3$

**(i) Define the term 'empirical formula' and deduce the empirical formula of hexane.** (3 marks)

**(ii) Define the term 'structural isomerism' and state the type of structural isomerism shown by hexane and 2-methylpentane.** (3 marks)

**(iii) State and explain which compound, hexane or 2-methylpentane, has the lower boiling point.** (3 marks)

**(iv) Draw and name another *two* structural isomers of $C_6H_{14}$.** (4 marks)

**(b) The structures of four isomeric alkenes are shown below:**

Compound A

$H_2C{=}CH—CH_2—CH_2—CH_3$

Compound B

$(CH_3)(H)C{=}C(H)(CH_2—CH_3)$

Compound C

$H_2C{=}C(CH_3)(CH_2—CH_3)$

Compound D

$(CH_3)(H)C{=}C(CH_3)(CH_3)$

**(i) Compounds A to D belong to the same homologous series. Explain the term 'homologous series'.** (3 marks)

**(ii) State the molecular formula of compound A.** (1 mark)

**(iii) State the empirical formula of compound B.** (1 mark)

**Total: 18 marks**

■ ■ ■

## Grade-A answer to Question 11

**(a) (i)** The empirical formula is the simplest whole number ratio ✓ of atoms of each element in a compound ✓. The molecular formula is $C_6H_{14}$, so the empirical formula is $C_3H_7$ ✓.

*e* The formula $C_3H_7$ cannot be simplified any further. It must be a whole number ratio.

**(ii)** Structural isomerism occurs when the molecules have the same molecular formula ✓ but a different structural formula ✓. This type of isomerism is called chain isomerism ✓.

*e* This type of isomerism occurs when there are two or more ways that the carbon skeleton of a molecule can be arranged.

**(iii)** 2-methylpentane ✓

The molecule is more branched and so has a lower surface area ✓ and weaker induced dipole–dipole forces between the molecules ✓.

*e* Remember: the more branched the alkane, the lower is its boiling point.

**(iv)**

```
              CH3                          CH3
              |                 ✓          |
CH3—CH2—CH—CH2—CH3             CH3—C—CH2—CH3 ✓
                                           |
      3-methylpentane ✓                    CH3    2,2-dimethylbutane ✓
```

*e* Another possible isomer exists and is an alternative correct answer:

```
      CH3   CH3
      |     |     ✓
CH3—CH—CH—CH3        2,3-dimethylbutane ✓
```

**(b) (i)** A homologous series is a group of organic compounds with the same general formula ✓. They have similar chemical properties ✓ because they possess the same functional group ✓. Their physical properties gradually change ✓ (e.g. boiling point increases) as the series is ascended, because the molecules possess an extra $CH_2$ ✓ group.

*e* There are 5 scoring points here; you need to score 3 to achieve full marks.

**(ii)** $C_5H_{10}$ ✓

*e* All the alkenes shown here are isomers, so they all have the same molecular formula.

**(iii)** $CH_2$ ✓

*e* The ratio of the atoms is 5:10. This simplifies to 1:2, so the empirical formula is $CH_2$.

# Question 12

# Alkanes

**(a) An alkane present in the kerosine fraction obtained from the fractional distillation of crude oil has a relative molecular mass of 198.0. This long-chain alkane undergoes thermal cracking to produce octane, $C_8H_{18}$, and propene, $C_3H_6$.**

**(i) Deduce the formula of the alkane.** (1 mark)

**(ii) Write an equation for the cracking reaction.** (1 mark)

**(b) Octane, $C_8H_{18}$, is a major constituent of the petrol fraction. Explain, with the aid of equations, how the combustion of octane in the internal combustion engine leads to the formation of the pollutant gases carbon monoxide and nitrogen monoxide. Explain, with equations, how a catalytic converter removes these two pollutant gases from the exhaust pipes of car engines.** (10 marks)

**(c) Butane and methylpropane are isomers with the molecular formula $C_4H_{10}$.**

**(i) Draw the structures of butane and methylpropane.** (2 marks)

**(ii) Write an equation for the complete combustion of methylpropane.** (1 mark)

**Total: 15 marks**

■ ■ ■

## Grade-A answer to Question 12

**(a) (i)** $C_{14}H_{30}$ ✓

The most common mistake is to add the two formulae together, producing an alkane with the formula $C_{11}H_{24}$. However, $C_{11}H_{24}$ has an $M_r$ of only 156.0. The difference between 198.0 and 156.0 is 42.0, which is equivalent to $C_3H_6$, so the formula is $C_{14}H_{30}$.

**(ii)** $C_{14}H_{30} \rightarrow C_8H_{18} + 2C_3H_6$ ✓

The equation must show the production of two propene molecules.

**(b)** Octane undergoes incomplete combustion ✓ to produce carbon monoxide and water:

$$C_8H_{18} + 8\tfrac{1}{2}O_2 \rightarrow 8CO + 9H_2O$$ ✓

The nitrogen present in the air also reacts with oxygen ✓ because the engine sparks ✓, producing the high temperature needed to overcome the energy of activation ✓.

$$N_2 + O_2 \rightarrow 2NO$$ ✓

The catalytic converter is a honeycombed ceramic structure supporting a platinum/rhodium catalyst ✓. Carbon monoxide is converted to carbon dioxide ✓; nitrogen monoxide is converted to nitrogen gas ✓:

$$2NO + 2CO \rightarrow N_2 + 2CO_2$$ ✓

Unburnt hydrocarbons also react with the nitrogen monoxide to produce carbon dioxide, nitrogen and water ✓:

$$C_8H_{18} + 25NO \rightarrow 8CO_2 + 9H_2O + 12\tfrac{1}{2}N_2$$ ✓

There are 12 scoring points for this part question; a maximum of 10 marks can be awarded. The answers given will vary, so the mark scheme is flexible. It is important that you learn the equations for the production and removal of the pollutants.

**(c) (i)**

$CH_3—CH_2—CH_2—CH_3$ ✓

Butane

$CH_3—CH(CH_3)—CH_3$ ✓

Methylpropane

Most candidates will draw the structures correctly.

**(ii)** $C_4H_{10} + 6\tfrac{1}{2}O_2 \rightarrow 4CO_2 + 5H_2O$ ✓

The combustion equation is quite often balanced incorrectly. Remember, complete combustion requires oxygen and involves the production of carbon dioxide and water. If there are four carbon atoms in the hydrocarbon, then four $CO_2$ molecules are produced. If there are ten hydrogen atoms in the hydrocarbon, then five $H_2O$ molecules are produced. Balance the equation by looking at the number of oxygen atoms. In this case there are 13 oxygen atoms on the right-hand side, so there must be $6\tfrac{1}{2}$ oxygen molecules on the left-hand side.